SURFACE TENSIONS

Manchester University Press

also available in the series

Utility reassessed
The role of ethics in the practice of design
JUDY ATTFIELD

Design and the modern magazine
EDITED BY JEREMY AYNSLEY AND KATE FORDE

The culture of fashion
A new history of fashionable dress
CHRISTOPHER BREWARD

Stained glass and the Victorian Gothic Revival
JIM CHESHIRE

The British country house in the eighteenth century
CHRISTOPHER CHRISTIE

'The autobiography of a nation'
The 1951 Festival of Britain
BECKY E. CONEKIN

Defining dress
Dress as object, meaning and identity
AMY DE LA HAYE AND ELIZABETH WILSON

The culture of craft Status and future
EDITED BY PETER DORMER

Representations of British motoring
DAVID JEREMIAH

Graphic design
Reproduction and representation since 1800
PAUL JOBLING AND DAVID CROWLEY

The Edwardian house
The middle-class home in Britain 1880–1914
HELEN C. LONG

The birth of modern London
The development and design of the city
ELIZABETH MCKELLAR

Interior design and identity
EDITED BY SUSIE MCKELLAR AND PENNY SPARKE

Dress and globalisation
Identity and the tactics of encounter
MARGARET MAYNARD

Establishing dress history
LOU TAYLOR

The study of dress history
LOU TAYLOR

STUDIES IN DESIGN

GENERAL EDITORS:
Christopher Breward and Glenn Adamson

FOUNDING EDITOR:
Paul Greenhalgh

Surface tensions

SURFACE, FINISH AND THE MEANING OF OBJECTS

Edited by Glenn Adamson and Victoria Kelley

Manchester University Press
Manchester and New York

distributed exclusively in the USA by Palgrave

Published by Manchester University Press
Oxford Road, Manchester M13 9NR, UK
and Room 400, 175 Fifth Avenue, New York, NY 10010, USA
www.manchesteruniversitypress.co.uk

Distributed in the United States exclusively by
Palgrave Macmillan, 175 Fifth Avenue, New York, NY 10010, USA

Distributed in Canada exclusively by
UBC Press, University of British Columbia, 2029 West Mall,
Vancouver, BC, Canada V6T 1Z2

British Library Cataloguing-in-Publication Data
A catalogue record for this book is available from the British Library

Library of Congress Cataloging-in-Publication Data applied for

ISBN 978 0 7190 8751 6 *hardback*

First published 2013

Typeset in ITC Giovanni
by Carnegie Book Production, Lancaster
Printed in Great Britain
by TJ International, Padstow

Contents

List of illustrations vii
List of contributors xii
Acknowledgements xv

Introduction *Glenn Adamson and Victoria Kelley* 1
1 A superficial guide to the deeper meanings of surface *Victoria Kelley* 13
2 Surface as practice *Lesley Millar* 26

I TIME, WORK

3 Layering: materiality, time and touch in Japanese lacquer *Christine M. E. Guth* 34
4 Patina and the bourgeoisie: the appearance of the past in nineteenth-century Paris *Manuel Charpy* 45
5 A falsification of temporality: carpet distressing in Morocco *Myriem Naji* 60
6 Textiles and time: conservation and public response *Margaret Ponsonby* 72
7 Surface engagements in Astana *Victor Buchli* 84

II TEXTILES, SKIN

8 'No less dangerous to use than that of a Basilisk': skin and seventeenth-century English embroideries *Mary M. Brooks* 98
9 Colour, language and light in mid-nineteenth-century women's fashion *Charlotte Nicklas* 111
10 *Wifedressing*: designing femininity in 1950s American fashion *Rebecca Arnold* 123

11 Long coats, flowing fabrics: fashioning masculinity and desire in film and television *Sarah Gilligan* 134

12 Surface as material, material into surface: dialectic in Carol Christian Poell *Ulrich Lehmann* 147

III DEPTH, IM/PERMEABILITY

13 Surface recognition: light and reflection in mid-nineteenth-century drawing rooms *Alice Barnaby* 166

14 A theory of the paint-by-number surface *Stephen Knott* 181

15 The materiality of print: issues of artifice and integrity relating to the printed ceramic surface *Steve Brown* 193

16 Substance abuse: the postmodern surface *Glenn Adamson* 200

List of illustrations

Plates

Plates between pp. 112 *and* 113.

1 Frances and Ron Geesin, *Tri-Aura*. Photography Ron Geesin

2 Ieva Krumina, *Donation*. Image courtesy of the artist

3 Lise Bjorne Linnert, *Desconocida/Unknown*. Image courtesy of the artist

4 Liz Nilsson, *Letting Go*. Image courtesy of the artist

5 Masayo Ave, *i-Haptic touch*. Image courtesy of the artist

6 Naoko Yoshimoto. Photography Lesley Millar

7 Shelley Goldsmith, *Outbursts 3 & 4*, *Revealing Evidence*. Image courtesy of the artist

8 Utamaro, 'Hour of the Hare', from *The Twelve Hours in the Yoshiwara*, colour woodblock print, c. 1795. Trustees of the British Museum 1954,0410.0.13

9 Makie lacquer writing table, wood, covered with gold and silver, c. 1620. Victoria and Albert Museum, London, given by the Misses Alexander W.339–1916

10 Negoro ware tray, lacquered wood, fourteenth century. Yale University Art Gallery, Collection Peggy and Richard Danziger, LL.B.1963

11 Ritsuô (Ogawa Haritsu), *inrô* in the form of a temple gong, eighteenth century, lacquer imitating patinated metal. Victoria and Albert Museum, London, Pfungst Gift W. 339-1922

12 Red Drawing room at Uppark. Copyright NTPL/Nadia Mackenzie

13 Curtain in Saloon at Uppark. Copyright NTPL/Nadia Mackenzie

14 William Whipple's bedroom at the Moffatt-Ladd House. Photography

by Ralph Morang. Courtesy Moffatt-Ladd House and Garden, Portsmouth, New Hampshire

15 Early nineteenth-century curtains at the Moffatt-Ladd House. Photography by Bruce Alexander. Courtesy Moffatt-Ladd House and Garden, Portsmouth, New Hampshire

16 Dressed and undressed women (clockwise from top left): Bathsheba (detail, panel, Ashmolean WA 1947.191.310); Sarah (detail, box, Ashmolean WA 1947.191.315; note one section of the skirt is a reconstruction); Esther (detail, panel, Feller Collection F52); Susanna (detail, sampler, Feller Collection F203)

17 Allegorical, personified and imaginary women (left to right): Charity (detail, panel, Feller Collection F40); Africa (detail, panel, *Four Continents and Scene of Rural Life*, Feller Collection F475); Mermaid (detail, *Proclamation of the Young Solomon*, Ashmolean WA 1947.191.313)

18 Faces and décolletages (clockwise from top left): Pattern drawer's inked drawing, unfinished design (Feller Collection F389); Queen of Sheba (detail, panel, Feller Collection F400); Personification of the Sense of Smell (detail, mirror, Feller Collection F19); Fashionable lady (detail, cushion, Feller Collection F62)

19 Solomon and the Queen of Sheba (clockwise from top left): Queen of Sheba, detail of embroidery; Queen of Sheba, radiograph; Mermaid, radiograph; Mermaid, detail of embroidery (Ashmolean WA 1994.142)

Feller Collection images: all images are copyright M & E Feller, photography by R Holdsworth FRPS. Ashmolean Museum images: the Ashmolean Museum of Art and Archaeology. Thanks to Mark Norman, the Ashmolean Museum of Art and Archaeology, for radiography, and to Sonia O'Connor, University of Bradford, for the image manipulation.

20 Evening dress, England, c. 1865, silk. Copyright Photo SCALA, Florence. The Metropolitan Museum of Art, New York

21 Day dress, probably France, c. 1845, silk taffeta. Collection of the Kyoto Costume Institute, photography by Taishi Hirokawa

22 Fashion plate, 'Public Promenade Dress', *The Ladies' Cabinet* NS 6 (May 1841), watercolour and engraving on paper. Special Collection, St Peter's House Library, University of Brighton

23 Fashion plate, 'Opera Dress', *The Ladies' Cabinet* NS. 7 (February 1842), watercolour and engraving on paper. Copyright Victoria and Albert Museum, London

24 Anne Fogarty, Ensemble, fall/winter 1955–56. C.I.59.16.6ab, Costume Institute, Metropolitan Museum of Art, New York. © 2011. Image copyright the Metropolitan Museum of Art/Art Resource/Scala, Florence

25 Anne Fogarty, Advertisement, 1953. Museum at the Fashion Institute of Technology, New York. © The Museum at FIT

26 Anne Fogarty, Dress, 1950–54. 92.59.1ab, Museum at the Fashion Institute of Technology, New York. © The Museum at FIT

27 Anne Fogarty, Ensemble, fall/winter 1957–58. C.I.63.47.3ab, Costume Institute, Metropolitan Museum of Art, New York. © 2011. Image copyright The Metropolitan Museum of Art/Art Resource/Scala, Florence

28 Torchère candelabrum, George Bullock, 1816–18. Copyright Victoria and Albert Museum, London

29 *The Awakening Conscience*, 1853, William Holman Hunt. Copyright Tate Britain, London

30 *Ballet Dancers "1,2,3"*. Oil painting set, Craft Master Palmer Paint Inc., 1953. 'Paint-By-Number Collection, ca. 1934–1957', Archives Center, National Museum of American History, Smithsonian Institution, Washington

31 *Flight South*. Preparation sketches, samples and test runs, Palmer Pann Corp., 1955. 'Paint-By-Number Collection, ca. 1934–1957', Archives Center, National Museum of American History, Smithsonian Institution, Washington

32 *Old Sadface*. Craft Master kit, 1969. Image courtesy of the Collection of Jeff McMillan (2009)

33 *Mandrake*, 2009, porcelain with screenprinted in-mould decoration. Photography Steve Brown

34 *Arlecchino the Brave*, 2009, porcelain with screenprinted in-mould decoration. Photography Steve Brown

35 Alessandro Mendini, burning of Lassù chair, for *Casabella* magazine, 1974. Courtesy of Atelier Mendini

36 April Greiman in collaboration with Jayme Odgers, cover for *WET* magazine, 1979. Copyright April Greiman and Jayme Odgers

37 Karole Armitage (choreographer), *Drastic Classicism*, 1981. Courtesy of Armitage Gone! Dance Company. Photograph by Terry Stevenson

38 Alessandro Mendini, *Proust Chair*, 1978. Wood and painted textile. Private collection

39 Ettore Sottsass, Cabinet No. 8, 1994. Yellow-lacquered maple wood with waterstain grain, natural Burled Elm. Edition of 6. Courtesy of Friedman Benda and the estate of the artist. Photography Jay Hanna

Figures

1 Funerary monument of Ralph of Shrewsbury, Bishop of Bath and Wells., c. 1363. Courtesy of Wells Cathedral. Photography copyright Stephen J. Gledhill *page* 3

2 Woman's embroidered linen smock, c. 1630. Copyright Victoria and Albert Museum, London 4

3 Desk-and-bookcase, Boston, Massachusetts, 1760–70, with disguised repairs dating c. 1956. Mahogany with white pine, Chipstone Foundation. Photography Gavin Ashworth 5

4 Arras Memorial and Faubourg D'Amiens Cemetery, Arras; Beaumont Hamel Newfoundland Memorial Park, the Somme. Photography Robert Sinfield 6

5 Bill Brandt, 'The Perfect Parlourmaid', *Picture Post*, 29 July 1939. Copyright Bill Brandt/Bill Brandt Archive Ltd 8

6 Hussein Chalayan, Central St Martins Graduation Collection, 1993. Copyright Hussein Chalayan 9

7 Marek Cecula, *Burn Again CH.3* from the *In Dust Real* series, 2005, industrial porcelain, wood fire. Photography Sebastian Zimmer 10

8 Fiona Banner, *Jaguar*, 2010, polished Sepecat Jaguar aircraft (8.69 × 4.92 × 16.83 m), as installed at Tate Britain. Courtesy of the artist and Frith Street Gallery, London 11

9 From a drawing by A. Brun, auction of antique goods at Hotel Drouot, with a bourgeois audience, c. 1880. Private collection of Manuel Charpy 47

10 Hotel Drouot and antique dealers' shops in Paris (1861 and 1901). Copyright Jean-Luc Pinol and Manuel Charpy 48

11 Antique dealer's shop in Grenelle Street, c. 1890. Archives de Paris, Fonds Union photographique française, UPF 1348 49

12 Gaston Noury, bourgeois women in an antique dealer's shop on banks of the Seine, watercolour, 1897. Private collection of Manuel Charpy 50

13 The shop of antique dealer Eugène Boban, boulevard Saint-Germain, Paris, c. 1870–80. Fonds Eugène Boban, Bibliothèque nationale de France, NAF 21476, pièce n° 275 50

14 Émile Bayard, *L'Art de reconnaître les fraudes*, Paris: R. Roger et F. Chernoviz, 1914. Private collection of Manuel Charpy 53

15 'Fields' of carpets surrounding a village where middlemen specialize in the washing and distressing of carpets. Photography Myriem Naji 62

16 A card (centre) used by weavers to create a nap on a brocaded carpet. Photography Myriem Naji 62

17 Detail of a tapestry woven carpet. Photography Myriem Naji 63

18 The surface of the brocaded black carpet is burnt with a blowtorch. Photography Myriem Naji 64

19 The carpet's surface is scraped both horizontally and vertically with a heavy-duty iron scraper. Photography Myriem Naji 65

20 'Cardboard architecture', Astana. Photography Victor Buchli 87

21 Old and new in contrast, Astana. Photography Victor Buchli 90

22 New cladding on Soviet-era buildings, Astana. Photography Victor Buchli 91

23 Prestizh, Astana. Photography Victor Buchli 92

24 Robert Downey Jr as Holmes in *Sherlock Holmes* (2009), directed by Guy Ritchie, produced by Bruce Berman. Image: DVD still 135

25 *Sherlock* (BBC 2010. 1:1), television series produced by Mark Gatiss, Steven Moffat and Sue Vertue. Image: DVD still 136

26 *Sherlock Holmes* (2009), directed by Guy Ritchie, produced by Bruce Berman. Image: DVD still 138

27 *Sherlock* (BBC 2010. 1:2), television series produced by Mark Gatiss, Steven Moffat and Sue Vertue. Image: DVD still 142

28 Carol Christian Poell, bull's blood-dyed leather trousers fitted front to back, 1999. Photography Stefan Ziesler. Copyright Carol Christian Poell 154

29 Carol Christian Poell, jacket of kangaroo leather, 2007. Photography Stefan Ziesler. Copyright Carol Christian Poell 156

30 Carol Christian Poell, bodyplast long skirt, silk, leather socks with flat sole, 2000, photography Steve Seleska. Copyright Carol Christian Poell. 158

31 Carol Christian Poell, transparent leather gloves, 2001. Photography Stefan Ziesler. Copyright Carol Christian Poell 159

32 Carol Christian Poell, silicone leather stick-on dress, stretch leather toe-stockings with mid-heel sole, 2000. Photography Steve Seleska. Copyright Carol Christian Poell 161

Contributors

Dr Glenn Adamson is Head of Research at the Victoria and Albert Museum and tutor at the Royal College of Art. Dr Adamson is co-editor of the triannual *Journal of Modern Craft*, and the author of *Thinking Through Craft* (Berg/V&A Publications, 2007), an anthology entitled *The Craft Reader* (Berg, 2010), and *The Invention of Craft* (Bloomsbury/V&A Publications, 2013). His other publications include the co-edited volume *Global Design History* (Routledge, 2011). He was, with Jane Pavitt, the co-curator for the exhibition *Postmodernism: Style and Subversion, 1970 to 1990*, which was on view at the V&A from September 2011 to January 2012.

Dr Rebecca Arnold is Oak Foundation Lecturer in the History of Dress and Textiles at the Courtauld Institute of Art in London. Her recent publications include two books, *The American Look: Fashion, Sportswear and the Image of Women in 1930s and 1940s New York* (I. B. Tauris, 2009) and *Fashion: A Very Short Introduction* (Oxford University Press, 2009).

Dr Alice Barnaby is a lecturer in English literature at the University of Bedfordshire. Her publications include: 'Dresses and Drapery: Female Self-Fashioning in Muslin, 1800–50', in *Artistry and Industry: Representations of Creative Labour in Literature and the Visual Arts c.1830–1900* eds. Patricia Zakreski and Kyriaki Hadjaifxendi (Ashgate, forthcoming, 2013); and 'Top-lighting in exhibition spaces: 1780–1860', in M. Henning (ed.), *Museum Media: The Handbook of Museum Studies* (Wiley-Blackwell, forthcoming 2013).

Dr Mary M Brooks is a consultant specializing in museums, conservation and textiles. She is a lecturer at Durham University and Honorary Research Fellow at the University of Glasgow. She is Joint Editor of *Textile History*.

Dr Steve Brown completed his doctorate at the Royal College of Art in 2011, a Collaborative Doctoral Award in association with the Victoria and Albert Museum and funded by the Arts and Humanities Research Council. He is a visiting lecturer at a number of UK and international universities including the Royal College of Art. Recent papers include 'The role of interpretative mediation in applied art screenprinting: a model for production' ('Impact 6': Interdisciplinary Print Conference, Centre for Fine Print Research, University of the West of England, 2009) and 'Lo-tech accessible approaches for 3D printing in glass' ('Crossovers': BECon Glass Conference, Portland, Oregon, 2011).

Dr Victor Buchli is Reader in Material Culture in the Department of Anthropology, University College London. Some of his previous books include *An Archaeology of Socialism* (Berg, 1999), *Archaeologies of the Contemporary Past* (Routledge, 2001) (with Gavin Lucas) and *Urban Life in Post Soviet Asia* (Routledge, 2007) (with C. Alexander and C. Humphrey). Victor is also a founding and managing editor of *Home Cultures*.

Dr Manuel Charpy is an alumnus of the École Normale Supérieure (ENS), holder of the *agrégation* (applied arts) and a doctor's degree in history, and is a researcher in CNRS (Centre National de la Recherche Scientifique) in IRHIS-University of Lille III. He works in particular on material culture, visual culture and urban culture during the nineteenth century.

Dr Sarah Gilligan is a lecturer in media at Hartlepool College, UK. Her recent publications include 'Fragmenting the black male body' (*Fashion Theory*, 2012) and 'Heaving cleavages and fantastic frock coats: gender fluidity, celebrity and tactile transmediality in contemporary costume cinema' (*Film, Fashion and Consumption*, 2011). She is currently writing a monograph entitled *Fashion and Film: Gender, Costume and Stardom in Contemporary Cinema* (Berg, forthcoming).

Dr Christine Guth leads the Asian design history and material culture specialism in the Royal College of Art and Victoria and Albert Museum postgraduate History of Design programme. Her recent publications include 'Import substitution, innovation and the tea ceremony in fifteenth and sixteenth-century Japan', in *Global Design History*, edited by Sarah Teasley and Glenn Adamson (Routledge, 2011) and 'The local and the global: Hokusai's Great Wave in contemporary product design' (*Design Issues*, 2012).

Dr Victoria Kelley is Reader in Design History and Material Culture at the University for the Creative Arts, Rochester, UK, and senior lecturer in the School of Fashion and Textile Design at Central Saint Martins School of Arts and Design, London. Her recent publications include *Soap and Water: Cleanliness, Dirt and the Working Classes in Victorian and Edwardian Britain* (I. B. Tauris, 2010) and 'The interpretation of surface: boundaries, systems and their transgression in clothing and domestic textiles, c.1880–1939' (*Textile: The Journal of Cloth and Culture*, 2009).

Dr Stephen Knott graduated from an AHRC-funded Collaborative Doctoral Award at the Royal College of Art/Victoria and Albert Museum with his PhD entitled 'Amateur craft as a differential practice' in July 2012. He has been awarded the first Postdoctoral Fellowship in Modern Craft at the Crafts Study Centre in Farnham and teaches at Kingston University and the Royal College of Art. His publications include an article on railway modelling, 'Keen House: scale and the architecture of enthusiasts' (*Design and Culture*, 2012) and a forthcoming work entitled 'Design in the age of prosumption: the craft of design after the object' (*Design and Culture*, 2013).

Professor Ulrich Lehmann studied philosophy, sociology and modern languages in Frankfurt-on-Main, Paris and London. His research interests are divided into the history of material culture 1780–1900, especially in France and Germany, and studies of contemporary design and photography. His most recent publications are: 'Series and multiples', in *The Small Utopia: Ars Multiplicata* (Prada Foundation, 2012); 'Making as knowing: epistemology and technique in craft' (*Journal of Modern Craft*, 2012); 'Mode gegen Moderne' [Fashion v. the Modern], in *Mode: ein kulturwissenschaftlicher Grundriss* (L. Bieger, A. Reich, S. Rohr eds) (Fink, 2012); and 'Sampling and the materiality of sound', in *Postmodernism* (V&A Publications, 2011).

Professor Lesley Millar is Professor of Textile Culture and Director of the Anglo-Japanese Textile Research Centre at the University for the Creative Arts. She is an exhibition curator specializing in textiles and has been project director for seven major international touring exhibitions: *Revelation* (1996–98), *Textural Space* (2001), *Through the Surface* (2003–5), *21:21 – the Textile Vision of Reiko Sudo and NUNO* (2005–7), *Cloth and Culture NOW* (2008), *Cultex: Textiles as a Cross-cultural Language* (2009–11) and *Lost in Lace* (2011–12.) She is currently leading an EU project 'Transparent Boundaries' (2012–13) with partner organizations in Denmark, Greece and Italy. In 2008 she received the Japan Society Award for significant contribution to Anglo-Japanese relationships and in 2011 was appointed MBE for her contribution to Higher Education.

Dr Myriem Naji is a research fellow in the Department of Anthropology at University College London, UK. She curated an exhibition related to her research on weaving in Morocco, entitled *Weaving the Threads of Livelihood: The Aesthetic and Embodied Knowledge of Berber Weavers* at the Brunei Gallery, School of Oriental and African Studies in 2011. Her publications on textiles include a catalogue for the exhibition and 'Gender and materiality in-the-making: the manufacture of Sirwan femininities through weaving in Southern Morocco' (*Journal of Material Culture*, 2009).

Dr Charlotte Nicklas is lecturer in the history of art and design at the University of Brighton, UK, specialising in the history of dress and textiles. She received her PhD in 2010, also from the University of Brighton, for a thesis entitled 'Splendid hues: colour, dyes, everyday science, and women's fashion, 1840–1875'. She has published reviews in *Textile History* and *Fashion Theory*.

Dr Margaret Ponsonby is senior lecturer in history and media and cultural studies at the University of Wolverhampton, UK. Her recent publications include *The Single Homemaker and Material Culture in the Long Eighteenth Century* (with David Hussey) (Ashgate, 2012) and 'Textiles and time: reactions to aged and conserved textiles in historic houses open to the public in England and the USA' (*Textile History*, 2011).

Acknowledgements

This book is the final outcome of the Surface Tensions research network, convened in 2009 as a partnership between the University for the Creative Arts and the Victoria and Albert Museum. The editors are very grateful to UCA, the V&A, and the Royal College of Art for their generous support of the network, its events and this book, with special thanks to Professor Seymour Roworth Stokes, Professor Kerstin Mey and Simon Bliss at UCA, to the Research Department at the V&A, and to Professor Jeremy Aynsley and Jane Pavitt at the RCA. Victoria Kelley would also like to record her thanks to the Central Saint Martins College of Arts and Design research fund for support in the latter stages of the book's preparation.

Introduction

Glenn Adamson and Victoria Kelley

PHILOSOPHER of aesthetics Thomas Leddy describes how, in his field, there is 'an entire class of neglected properties'.[1] These are concerned 'not with the underlying structure or substance of things, but with what might be called surface properties'.[2] Skin deep; shallow; decorative; superficial: language like this tends to condition our instincts about surfaces. They are the external appearance of things, easily manipulated, and within many traditions of thought, are held to be of lesser consequence than 'deeper' or more 'substantive' interiorities – hence the neglect identified by Leddy, as commentators rush past the surface in their hurry to excavate more complex inner truths, and designers try not to draw attention to their work on surfaces.

Yet it is also possible to conceive of the surface in more positive, less absolute terms: as a site where complex forces meet. Surfaces can be theorized as membranes, protective shells, sensitive skins, even thicknesses in their own right (as anyone with a microscope can tell you, disentangling 'surface' from 'depth' can be a difficult task). The surface is not so much a barrier to content as a condition for its apprehension. Contemporary design increasingly challenges the surface/depth binary, partly through technological means, as new materials and digital processes allow designers to conceive and produce more complex objects and buildings. So there is good reason for attending to surface, as it becomes more complex and convoluted in its applications. Yet in historical practice, the surface/depth binary has been influential, and it is important that we attend to the effects of this asymmetry. Surface has often been overlooked in a quest for depth and 'truth'. This collection of essays consciously unsettles this dynamic, instead charting a model of a 'complex' surface based not in geometry or topology, but within the field of design history, a discipline that responds to contemporary concerns and practices while investigating the past with rigour. The current collection is thus able to make a timely recognition of 'surface' as a crucial quality for the apprehension of objects in both past and present, evidencing the historical preoccupation with the surface as the superficial opposite of depth or structure, as well as contemporary efforts to complicate its understanding.

Design history is also a field that actively explores disciplinary boundaries. So while the essays included here all pursue design history's central concern of paying attention to and understanding the material world of things, they may also relate to other disciplinary allegiances in the study of, for instance, textiles, fashion, film, philosophy, anthropology and contemporary creative practice. The range of chronologies addressed is wide (seventeenth century to the present day) as is the range of geographies (from Japan to the United States of America, and many points in between). What unites this disparate material, apart from the theme of surfaces themselves, is a range of concerns that are captured in the three sections of the book: Time, work; Textiles, skin; Depth, im/permeability. Time, work explores the particular characteristic of surface as the site of labour intensive qualities of finish and ornamentation, as well as the way it registers temporality, process and mutability in the breakdown or upkeep of these qualities. Textiles, skin recognizes the notable relationship of fashion and clothing to surface through appearance and style, both materially and ontologically. Depth, im/permeability investigates the complications of the surface/depth binary, and what happens when a surface breaks down, or the binary is overturned. These categories are useful organizationally, but they are not, of course, meant to be monolithic or exclusive, and themes carry over from section to section.

It is customary in anthologies like this one for the editors to briefly summarize each essay in the introduction. We have elected not to do this here – hopefully our readers will have the time and interest to read each of the contributions (if not, Chapter 1 briefly cites each essay in surveying the theoretical and contextual literature around the issue of surface). Instead, in this introduction we offer our own readings of eight objects that will not be found elsewhere in the book. Our intention has been to show the range of circumstances in which surface comes to the fore (literally) as the bearer of meaning. Although they differ widely in date, material and genre, all of these examples show the complexity of the surface as a site. Temporality, intention, skill and concept are inscribed into each of these examples. Physically, this rich content is only skin deep; but that certainly does not lessen the complexity of reading that it demands. We suggest only a few cursory ideas that might adhere to each of these objects – hopefully, in so doing we also suggest the many directions in which examination of the surface from an interdisciplinary perspective might take us. These are the many paths across the surface that the rest of this book aims to pursue.

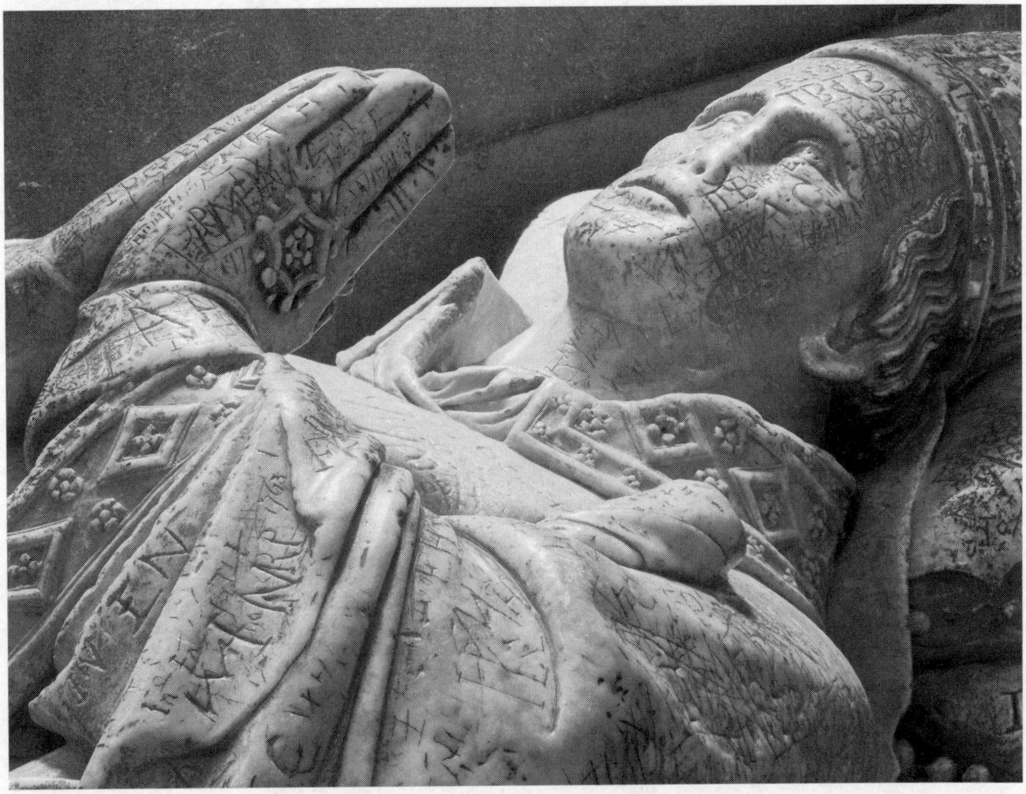

1 Funerary monument of Ralph of Shrewsbury, Bishop of Bath and Wells, c. 1363, Wells Cathedral

This funerary monument, a life-sized portrait of a bishop, was once only a fine example of medieval sculpture. But the surface exhibits extensive wear, including countless names, initials and dates that have been carved into the stone over the centuries. This accumulation of marks initially seems shocking, an act of defacement or even desecration. Yet to most contemporary viewers, the poignancy of the object has probably been dramatically increased rather than diminished by this rough treatment. Typically surfaces register the passage of time only approximately – as in the softening of the bishop's facial features, which have been worn away by the touch of countless hands. Here this slow, human-induced erosion combines with the very specific moments recorded in the graffiti and the original fourteenth-century intention to fix a great man permanently in stone, forming a rich palimpsest of materialized temporality.

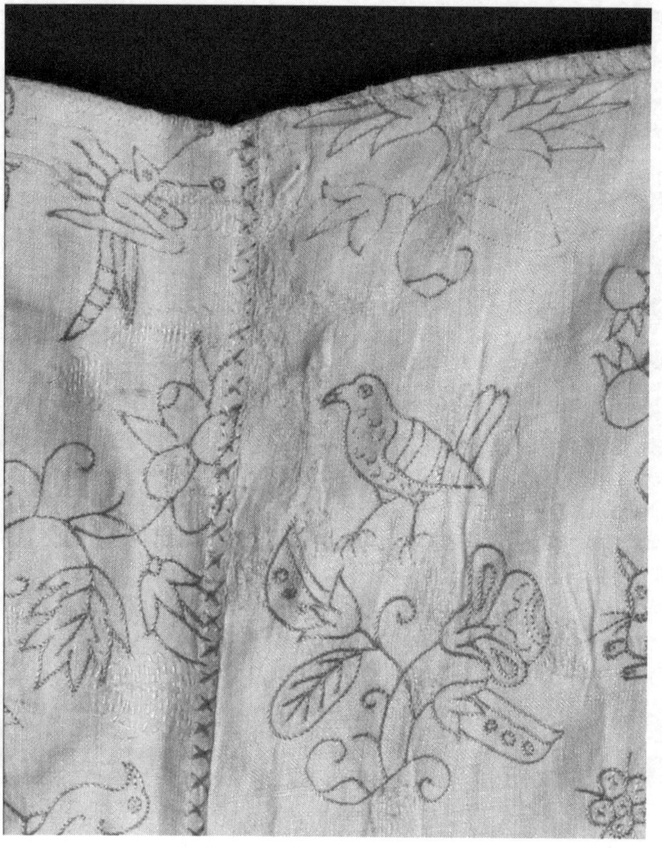

2 Woman's embroidered linen smock, c. 1630, Victoria and Albert Museum, London

This garment, from the Victoria and Albert Museum, is a rare survival. Many museum dress exhibits, precisely because they are beautifully conserved and displayed, lack an essential sense that they have been worn, that they are historic objects with a past that has brought them into intimate contact with human bodies. This garment displays its history quite clearly, with several patches of discoloration, and a delicately executed darn that reinforces and makes good an area of worn cloth. Clothing is the surface that the body presents to the public gaze (and as well as being a surface, it also *has* a surface, embellished with embroidery and finished with starching and pressing). In this instance the surface can be seen as the site of interaction and ongoing processes of maintenance.

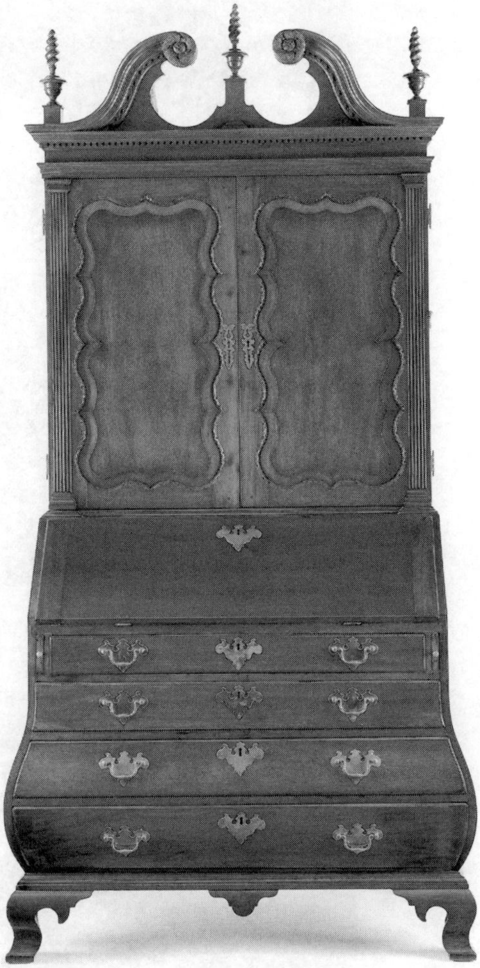

3 Desk-and-bookcase, Boston, Massachusetts, 1760–70, with disguised repairs dating c. 1956. Mahogany with white pine, Chipstone Foundation

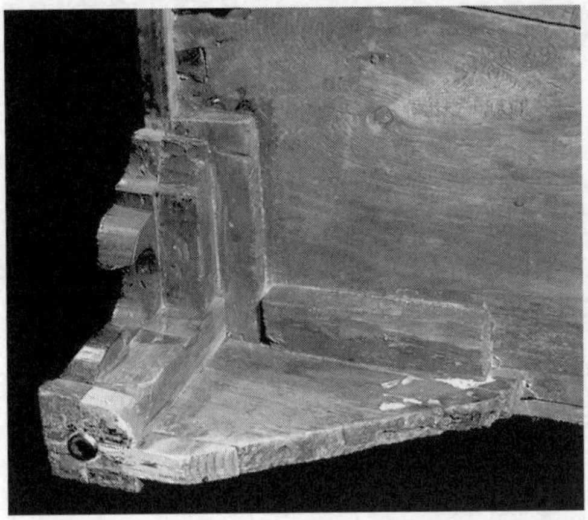

Furniture making is a trade intensely concerned with surface, as the diversity and complexity of finishing techniques suggests. For this reason, surface is also a key concern for fakers. This *bombé* desk-and-bookcase is partly original, but includes many disguised adaptations, introduced by a 1950s antique dealer who wished to pass the object off as entirely authentic. It originally had mirrors mounted on the doors, for example. A detail photograph taken from below shows how the feet – which are modern fabrications – disguise corner blocks of inconsistent colour. In other cases a surface may be deliberately distressed to make it look old, with scratches or ink and water stains introduced to mimic wear. The repetitive character of the marks sometimes gives such tactics away. The methods of connoisseurial analysis are, more often than not, based on the skills of reading a surface.

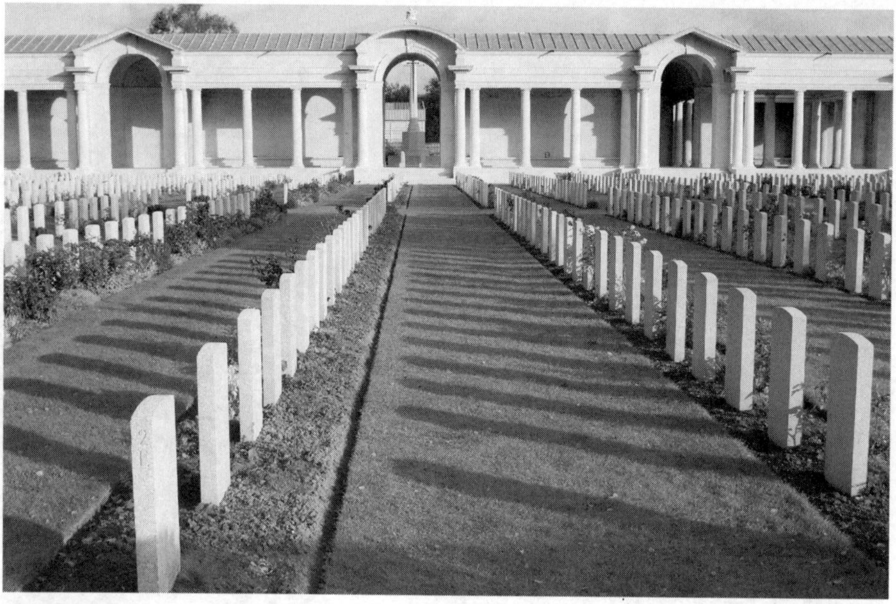

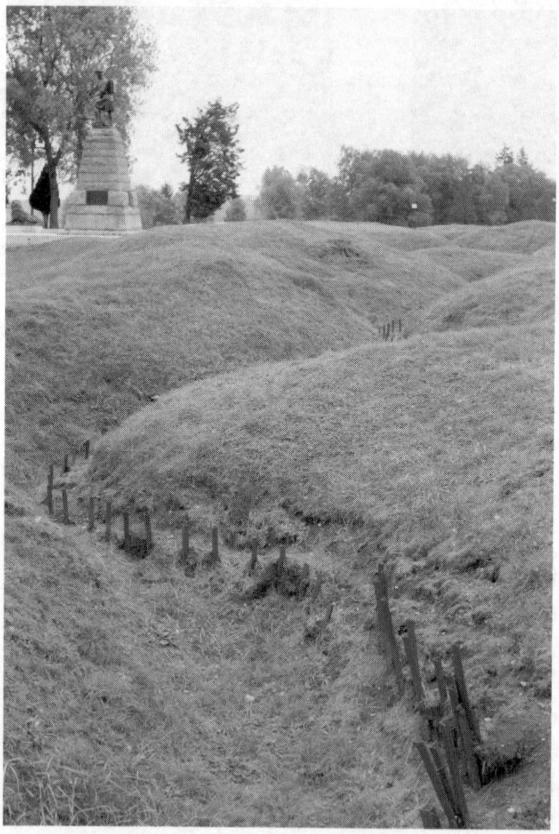

4 Arras Memorial and Faubourg D'Amiens Cemetery, Arras; Beaumont Hamel Newfoundland Memorial Park, the Somme

The First World War cemeteries of Belgium and northern France are very orderly, very peaceful places. The graves are kept in good repair, amid flowerbeds and immaculate green lawns. In many respects it seems very right that these cemeteries are such tranquil and well-organized places. What is so revolting about war is its radical lack of order, in the destruction of townscapes and landscapes, and, most shockingly of all, in the bodies of those killed and mutilated. In the cemeteries order is reimposed upon disorder, as the trenches and craters and scars of war have been carefully replaced with neat and calm serenity. This is also the case in the landscape around, where, in the years after the war, displaced populations returned and laboriously reclaimed the fields, filling in trenches, removing an 'iron harvest' of shell cases and barbed wire, and rededicating the land to cultivation. Over time the ploughs and harrows have gradually restored the smooth rolling swell of the topography. A domestic analogy springs to mind, that of *polish*: confronted with a deeply scratched or scored polished surface, a fitting response might be to sand it down and repolish it, to work over the scores and scratches. But in a few places in Belgium and France the trenches and shell holes remain, as gradually softening scars. They have been deliberately preserved, incorporated into memorial landscapes that are mournful, evocative and radically deny the impulse to re-order and to smooth over. These memorial parks with their still-inscribed trench landscape are a testimony to what happened. Here the trauma has not been smoothed away, the surface of the landscape has not been repolished and made good.

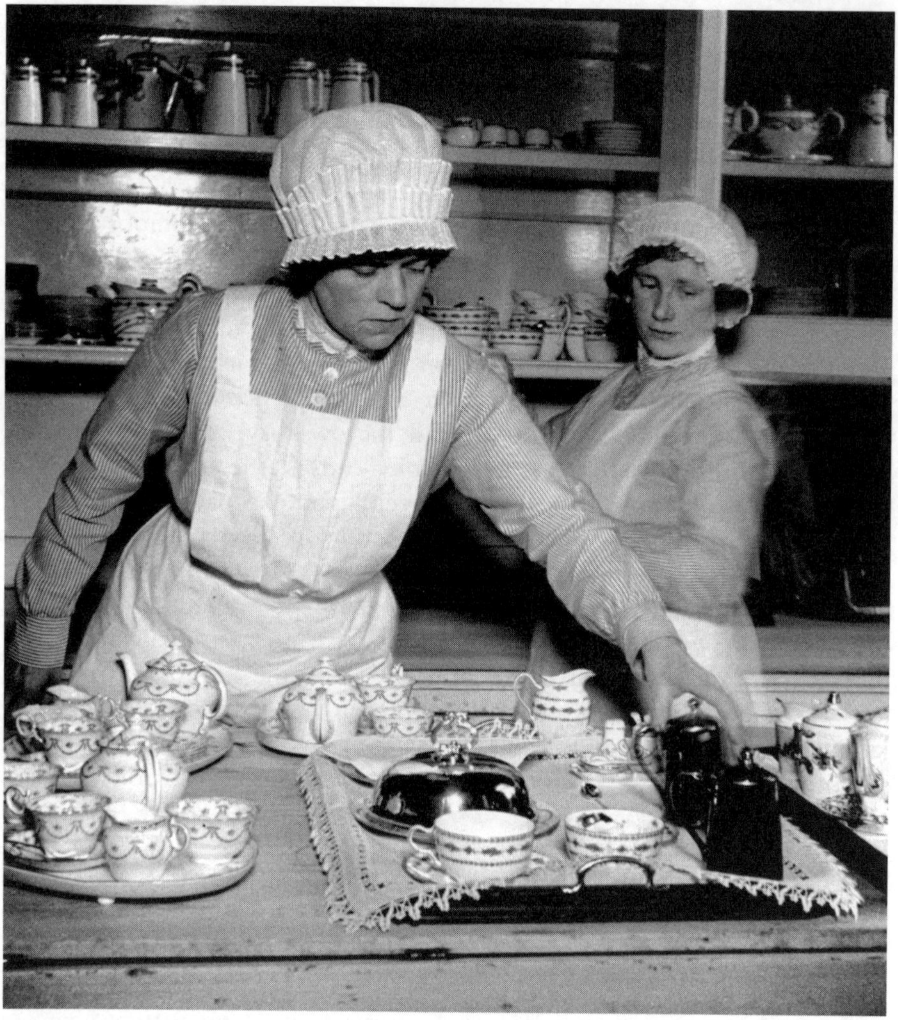

5 Bill Brandt, 'The Perfect Parlourmaid', *Picture Post*, 29 July 1939

Brandt's photographs of 1930s British social *milieux* include many images of the servants in his uncle's house. These images document the attention paid to surfaces in such a social setting (polished silver, starched table linen, garments that are meticulously pressed and 'goffered' to achieve a smooth, crisp finish and precisely defined frills). The strategies chosen by Brandt in photographing and printing suggest strongly that he was making a conscious effort to capture these material properties, that he recognized them as significant and socially articulate – redolent, perhaps, of prosperous respectability.

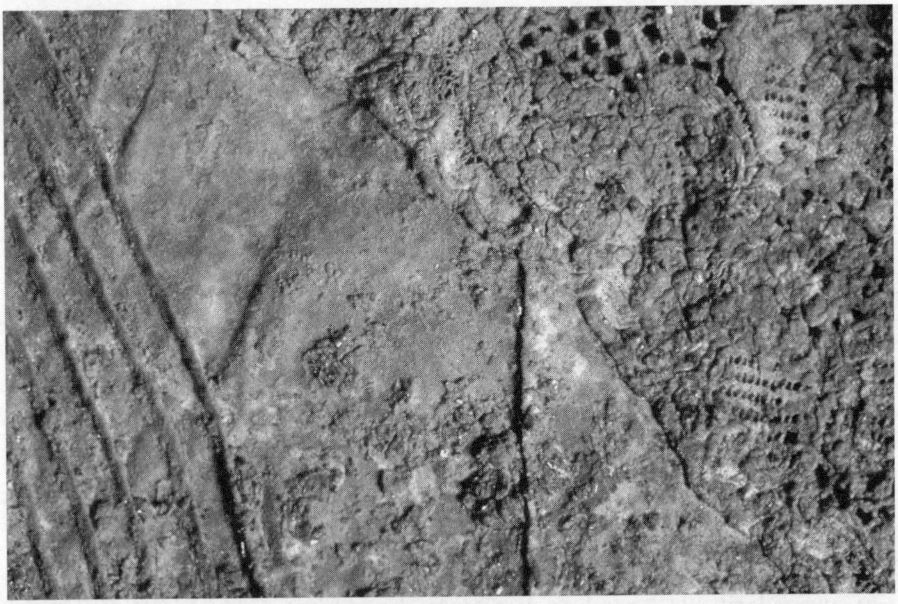

6 Hussein Chalayan, Central St Martins Graduation Collection, 1993

Hussein Chalayan took the garments of his final BA collection, and, some weeks before assessment, buried them in a friend's back garden, with iron filings sprinkled between the layers of cloth. The result, on disinterment, was fabric richly encrusted with a surface layer of rust, suggestive of great age and decay, despite being achieved quickly and artificially. Chalayan's concerns coincide with those of many other textile and fashion designers who have played with the expressive qualities of surface, producing effects that are fabricated 'quotations' of the past. By this means the surface effects of wear and tear and dirt, signs of poverty, age and degradation, become sanitized in affluent society, nostalgically absorbed into the fashion system.

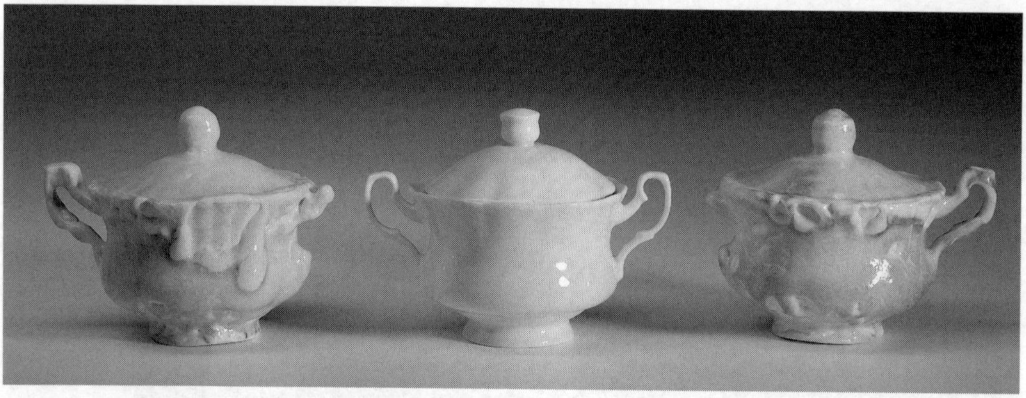

7 Marek Cecula, *Burn Again CH.3*, from the *In Dust Real* series, 2005, industrial porcelain, wood fire

In his *In Dust Real* series, the contemporary ceramic artist Marek Cecula makes new objects look like ancient remnants, or perhaps finds from a war-ravaged city. His method is simple: he subjects industrially produced porcelain to firing in a traditional wood-fuelled kiln, which deposits a surface of ash on the surface of the object. This re-firing of the objects transforms their clean perfection into a rich weathering. Sometimes fragile parts of the porcelain do not survive the process, creating an additional impression of time passing. Though many traditional ceramics (notably Japanese tea ceremony wares) are prized for their subtle wood-fired surfaces, in Cecula's work the effect is somewhat disconcerting, mimicking the passage of time and dramatizing, as he puts it, 'the fire as a creative and destructive force'.[3]

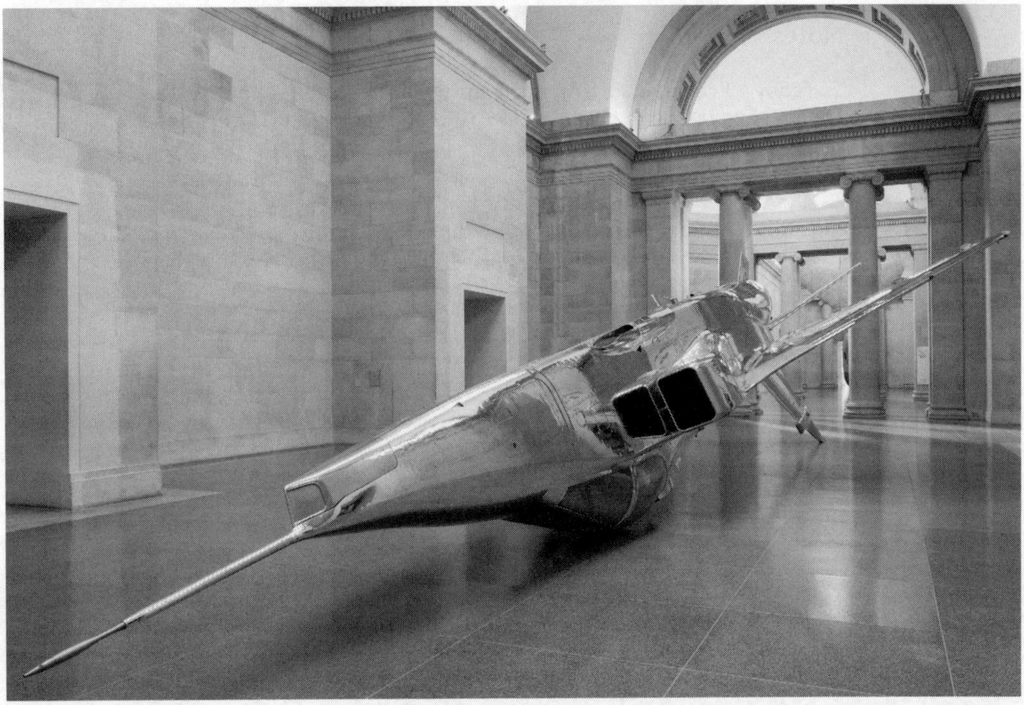

8 Fiona Banner, *Jaguar*, 2010, polished Sepecat Jaguar aircraft, as installed at Tate Britain

An unusually monumental essay in the tradition of the 'assisted readymade,' Fiona Banner's sculpture is exactly what it seems to be: a Sepecat Jaguar fighter aircraft which had seen action in Bosnia, stripped of its paint, polished by hand and recontextualized in an art gallery. This awe-inspiring found object can be read in many ways: as a way to confront the public with the usually remote weaponry that their government creates; an illustration of the overlap between technology and biology (the plane is named for a jungle cat and has the appearance of a bird of prey); or perhaps as a totem of uncontrolled aesthetic desire. Banner has said of military aircraft, 'We all hate war but these objects inspire a strange enthusiasm in us … They are absolutely designed to function and that function is to kill, and that says something questionable about our aesthetic judgement and makes us ask questions about our moral position.' Her decision to radically transform the surface of the fighter is key to its meaning. When you look at it, you find yourself reflected in it. The artwork is a giant distorted mirror of its own audience. In Banner's words, 'you can't detach yourself from it' – an apt slogan for any work that addresses the complexities of surface.[4]

Notes

1 T. Leddy, 'Everyday surface aesthetic qualities: "neat", "messy", "clean", "dirty"', *Journal of Aesthetics and Art Criticism*, 53:3 (Summer 1995), 259–268.

2 *Ibid.*

3 www.marekcecula.pl/sites/art_works11.php.

4 M. Brown, 'Tate Britain: Fiona Banner exhibition', *Guardian*, 28 June 2010, www.guardian.co.uk/artanddesign/2010/jun/28/tate-britain-fiona-banner, accessed 7 December 2012.

1 ✧ A superficial guide to the deeper meanings of surface

Victoria Kelley

Surface is often contrasted with what lies beneath or within, in a series of binary oppositions: surface/depth, surface/structure, surface/core. Such contrasts often suggest, at least in everyday language, that the real value or interest lies in the second half of the pairing, not the *superficial*, but the *in-depth* – but what might we find if we stop at the surface and consider it in material and metaphorical terms?

Surface is the topmost or outermost layer of an object or substance. It is the part of an object that is most accessible to our senses, often the only part that can be directly seen, or touched, at least at first encounter. Despite the fact that surface is frequently belittled or ignored in language and in discourse, in material practice surface layers are often given special attention, coated with substances or treated with processes that alter their appearance, including their colour and their feel (glaze on ceramics, varnish on wood, cosmetics on human skin) and that are two-dimensional applications upon three-dimensional objects. Thus the material complexity of the surface, as well as its assiduous formation, invites investigation. Many objects have not one but many surfaces, layered upon each other (clothed bodies, lacquered wood): indeed some objects, most notably textiles, are all surface, two-dimensional and planar, useful for wrapping, screening and veiling. Because they are impacted upon by the world around them, surfaces are usually subject to change over time (soiling, fading, creasing, cracking). As a result of this susceptibility, they are the site of strenuous routinized efforts to stave off or reverse such change, making them a terrain for many ongoing human interactions with the material world (cleaning, brushing, buffing, polishing) that are worth investigating for their social and symbolic meanings. *Boundary* is a word that often crops up in association with surface, suggesting the surface's role in defining, separating and containing, including and excluding. But boundaries are just as interesting when they break down as when they stand firm, so we might also consider what happens to surface when it is transgressed, or when the surface/depth binary does not hold, as is frequently the case. Finally, surface is almost always the site of representation (pencil on paper, paint on canvas, embroidered stitch on linen or silk, pixels on a screen), the context in which images are materially produced.

This chapter briefly surveys the existing historiography of the surface: not all the works cited state surface as their primary concern, but all can be used to cast light upon it. The chapter's structure anticipates the structure of this book, with three sections, on time and work; textile and skin; and depth and im/permeability. This categorical approach is inevitably both partial and provisional. It is impossible to render a term so various in its possible meanings, material and metaphorical, in a neat and comprehensive way. (And, it should be added, the word 'surface' is even more multifarious in its meanings should one stray into other academic disciplines than design and material culture – biology, physics, geometry and many other fields all have their own debates around 'surface'.) Thus if these three sections do not precisely form an introduction to the papers that follow, it is hoped that they will at least provide some useful contexts in which those investigations can be considered.

Time, work

Surface is the site of many interactions between people and things (as Lesley Millar's exploration of touch, 'Surface as practice', in Chapter 2 makes clear), including the labour-intensive and time-consuming processes of patterning and finishing (new objects) and maintaining (objects in use). Let us commence with the new, and those surface techniques that form the final stage in the production of many designed and crafted objects. The first and obvious point is to do with the status of the 'worked' surface in discourses around architecture and design. The decorated surface has long been contentious in design debates, and from the mid-nineteenth century onwards attempts were made to rationalize and reform the types of applied decoration used on the surfaces of buildings and (increasingly mass-produced) products, and then to do away with applied decoration altogether.[1] Similarly, applied surface finishes such as varnish and French polish were scorned and satirized, at least in their more excessive forms.[2] (Yet beyond the western context, this scorn for elaborate surface finishes did not arise, as Christine Guth shows in Chapter 3, her examination of the highly skilled and esteemed techniques of Japanese lacquer work).

Disapproval of the excessive surface became a commonplace of discourse around the creative arts. In 1908, Adolf Loos notoriously associated 'ornament' with 'crime'.[3] He saw surface elaboration as primitive, backward and anti-modern, critiquing the perceived irrationality of investing effort and skill in decorative effects that had no functional purpose. Under the 'yoke of ornament', Loos asserted, the 'ornamental craftsman has to work for twenty hours to reach the pay a modern worker earns in eight'.[4] (See Chapter 4, Manuel Charpy's exploration of the artificial ageing of fake 'antiques' in nineteenth-century Paris for an interesting take on the work of surface). Increasingly the design pioneers of the Modern Movement sought to remove surface elaboration from their buildings and products, pursuing instead 'technology, form, and especially space',[5] adhering to Le Corbusier's dictat of the whitewashed wall, as set out in *The Decorative Art of Today*: 'Law of Ripolin. Coat of whitewash. Elimination of the equivocal.'[6] In the 1920s and 1930s, buildings such as Le Corbusier's Villa Savoye embodied the Law of Ripolin, which arguably had its analogy in such designed objects as Mies Van Der Rohe's Barcelona Chair, that celebrated the smooth industrial surface of chrome-plated tubular steel.

The flat and unadorned modernist surface and the rhetoric around it has continued to provoke debate and discussion in architecture and design. Mohsen Mostafavi and David Leatherbarrow's essay *On Weathering* is a fascinating polemic on the relationship of architecture to time, with great emphasis on surface effects and surface integrity. Mostafavi and Leatherbarrow describe Le Corbusier's Villa Savoye as 'an X ray of beauty ... revealing the ... outline of things, their volume and color, as absolute, without the possibility of mistake' and they deliver a scathing and provocative critique of modernist buildings' inability to weather successfully, to depart from the absolutism of the white surface without fatally compromising their aesthetic effect.[7] Mark Wigley's work on the white walls of modern movement architecture also investigates the extent to which the removal of surface elaboration not only moved away from concern with surface towards concern with volume, space or mass but also, and rather contradictorily, represented an intensification of interest in surface, which was fetishized for its simplicity and 'truth', a process that Wigley argues had many parallels with the clothing fashions of the time.[8]

This prompts consideration of the gendered nature of surface: Paul Rudolph's brutalist Yale Art and Architecture Building, with its striated concrete surface that departed somewhat from the conventions of the material, was described by critics as 'effeminate' and even condemned as being 'dressed in ornamental drag'.[9] In modernist discourse, surface ornament was unequivocally gendered female, and attempts were made to displace or reform it.[10] In the 1920s and early 1930s the textile workshops of the Bauhaus minimized pattern and applied decoration (print) in favour of innovative weave techniques that were concerned with the structure of warp and weft rather than surface decoration (although the aesthetic effects achieved may ultimately have been decorative).[11] Yet Pat Kirkham has uncovered the persistence of decoration in much popular modernism, in for instance, the textile designs of Ray Eames, and the eclectic and exuberant interior schemes Eames constructed with her partner Charles.[12]

Architecture and design discourse has continued to be preoccupied with surface beyond the reaches of modernism. In the postmodern 'decorated shed' identified by Denise Scott Brown and Robert Venturi in *Learning from Las Vegas*, focus switches from form to surface, as the building becomes a sign, a billboard with communication as its ultimate purpose.[13] Postmodern buildings and objects also frequently feature artificially deconstructed surfaces, where the effects of time and duration are simulated, accompanying a range of other approaches that all challenge the integrity of the straightforwardly coherent building. (Chapter 5, Myriem Naji's account of the distressing of Moroccan carpets, is a contemporary example of deconstruction at the point of contact between two different systems of aesthetic value.)[14] This brings us to the surfaces of objects in use and over time. Surface is a pressing concern not just in the production of new artefacts, but also in ongoing relationships between things and people; time, as Mostafavi and Leatherbarrow note, is registered primarily on the surfaces of objects (and not just buildings) through *finish*.[15] And as the effects of time – dirt and stains, wear and tear – are situated on the surface, so too are the techniques of maintenance – washing, cleaning and polishing. Mary Douglas's classic anthropological account of symbolic beliefs around purity and pollution is of enormous value in understanding attitudes to cleanliness and dirt, tidying and maintaining.[16] Douglas works with structural binaries, and is concerned with ordered systems. Indeed, it is one of her great insights to show how attitudes to cleanliness and dirt are not the result of either irrational superstition, or mechanistic (or medical) materialism, but are instead founded upon creative attempts at ordering in the social sphere:

> In chasing dirt, in papering, decorating, tidying, we are not governed by anxiety to escape disease, but are positively re-ordering our environment, making it conform to an idea. There is nothing fearful or unreasoning in our dirt-avoidance: it is a creative movement, an attempt to relate form to function, to make unity of experience.[17]

Douglas is especially concerned with classifications, and with the boundaries that support them, and this makes her work of great interest in the consideration of surface; surfaces that are worn or torn or soiled strongly suggest broken boundaries and blurred classifications (thus, in the preserved interiors of historic houses, the matter of whether to restore and clean aged textiles and furnishings is an emotive one for curators and visitors, as Margaret Ponsonby shows in Chapter 6).

If Douglas's analysis is founded upon structuralism and its appreciation of systems, Gaston Bachelard's work is phenomenological.[18] In *The Poetics of Space*, Bachelard is concerned primarily with the experience of domesticity, which he accesses through literary evidence – the writings of poets, novelists and autobiographers whose imagery evokes the materiality and emotion of home. It may seem perverse to suggest an exploration of space as useful in the comprehension of surface, given the way in which surface is often opposed to structure and depth. Yet Bachelard describes surface as having expressive and emotive characteristics that help to constitute the notion of home. In discussing the polishing of furniture, Bachelard creates an analogy between 'polishing' and 'cherishing':

> From one object in a room to another, housewifely care weaves the ties that unite a very ancient past to the new epoch. The housewife awakens furniture that was asleep ... A house that shines from the care it receives appears to have been rebuilt from the inside; it is as though it were new inside.[19]

Bachelard's writing emphasizes the interiority of the home and the objects within it (the book's impetus is always inwards, to the intimate corners and spaces of the home). Thus interior as well as exterior surfaces are considered, notably in a passage based on a quotation from Jules Michelet that describes the construction of a bird's nest with its smooth lining: 'by exercising an active pressure, the female makes this into a felt-like padding ... modeled by fine touches, which make a surface originally bristling and composite into one that is smooth and soft'.[20]

Finally in this section it is worth mentioning the ideas of sociologist Stuart Walker, whose concern is with sustainability in contemporary design. Walker proposes innovative approaches to surface in order to maintain 'aesthetic longevity' in consumer products.[21] He argues that many modern manufacturing techniques produce objects with an aesthetic appeal that is largely based on 'shiny, highly polished, new surfaces'. These surfaces are vulnerable to degradation: 'monochromatic, glossy surfaces on painted metals and moulded plastics are delicate and highly susceptible to marring'. Walker suggests reusing materials, and employing techniques of 'rough work' in which fine surface is not part of the aesthetic appeal. 'One more scratch on a variegated, irregular surface that is an integral part of the object's design will be unlikely to cause aesthetic dissatisfaction.'[22] Grant McCracken's 1988 account theorized patina historically as a mark of status in inherited objects.[23] Walker reasserts patina as a strategy in sustainability, a desirable quality of products that consumers engage with over prolonged periods of time, an antidote to the unsustainability of throwaway consumer culture. (For an interesting take on 'throwaway' artefacts, see Chapter 7, Victor Buchli's analysis of urban development in Astana, where spectacular new buildings begin to decay even before they are completed.)

Textiles, skin

The surface that most people present to the world around them is composed, to a greater or lesser extent, of their clothing. It is no accident that clothing and fashion are often condemned as superficial, or at least potentially so – layers of disguise that may conceal or reveal the identity of the person underneath, a façade. 'All fur coat and no knickers', my mother's friends used to say about women who they believed to have an excess of style and a deficit of (economic or moral) substance. Oscar Wilde proved the conventional power of the notion by turning it, in part, upon its head: 'It is only shallow people who do not judge by appearances.'[24] So, if we are to talk about surface, we must talk about textiles, and fashion, and we should also mention the related category of skin. The relationship of body, skin and clothing (the 'second skin') has been much remarked upon by historians and theorists of fashion. This can be about simple proximity, as Claire Pajaczkowska argues:

> As cloth in clothing is the most tactile of surfaces, always in contact with skin and body, it carries the contradictory meanings of being an external surface turned outwards towards the gaze of the viewer, while remaining forever proximate.[25]

Pennina Barnett provides an etymology of the word 'cloth' which she suggests gives it the meaning of 'that which clings to the body'.[26] But the association can go further: Georges Vigarello's analysis of the practice of cleanliness in France notes how in the seventeenth century being clean was not a question of regular immersion in hot soapy water, but rather of an elaborate system of brushing and wiping the body and hair, and regularly changing white linen undergarments which soaked up the excretions of the body. As these white shirts and shifts spilled out of heavy over-garments at wrist and neck, they came to be seen as representative of clean skin, standing in for it.[27] Kitty Hauser's essay on 'The fingerprint of the second skin' is a detailed exploration of the analogous relationship between wearers and their denim jeans. The seams and stress-points of the jeans develop patterns of creasing, fading and abrasion that are akin to fingerprints and reveal the habits of the wearer, as well as the actions of the anonymous workers who sewed the garments.[28] (In Chapter 8, Mary Brooks explores the representation of skin by textile in seventeenth-century embroideries.)

Textiles and clothing are arguably a rather special category of surface because, as exemplified in Hauser's jeans that both screen the bodies of their wearers and lose their surface integrity in everyday wear and tear, they both *are* surface and *have* surface. A piece of cloth is a sheet or membrane in which an object or a body can be wrapped, more or less closely, to provide it with an outer surface. Yet a piece of cloth is also a substrate upon which surface elaboration (embroidery, print) can be applied. The surface decoration of textiles was well established as a supposedly feminine trait by the late nineteenth century. Decorative needlework skills in particular were feminized, and thus, as Bridget Elliot and Janice Helland argue, also 'marginalized and trivialized'.[29] Decoration, gendered as female (and also as 'primitive', as we have seen in Loos's analysis), has been identified by Tag Gronberg as 'Modernism's other'.[30] (See Chapters 9, 10 and 11, by Charlotte Nicklas, Rebecca Arnold and Sarah Gilligan, for explorations of the relationship of gender to textile surface.) But if the surface of cloth can be decorated it can also be subject to other, elaborate yet subtle surface processes. Examples are the fulling of woollen cloth, and the 'beetling' of linen, a process of beating with wooden hammers that tightens and burnishes the surface.[31] This recalls Gaston Bachelard's smooth, felted interior of the bird's nest, a surface that is *finished*: it has been assiduously worked upon to produce a particularly close and dense effect that neatly bounds the composite substrate beneath.

This is the *perfect* surface, and it is not unique to textiles; smooth and tight integrity is arguably a common surface characteristic across many materials, object types and historical periods. But what of surface imperfections? If time, as Mostafavi and Leatherbarrow argue, is registered upon the surfaces of buildings, then it is also registered upon the surfaces of textiles, and indeed, skin. The perfect surface has been challenged in textiles and fashion as in architecture and design, and surface imperfections have been artificially simulated, rather than left to develop naturally. The practice of deconstruction, as inspired by Jacques Derrida, implies the dismantling of binaries and the complication of monolithic truths and either/or answers. Derrida's work was both influential in and provocative to the formation of postmodern thinking, and has been used, beyond the philosophical sphere, to suggest material strategies in creative practice. Architecture took on deconstruction enthusiastically, as briefly noted in the previous section. Creative practice in fashion and textiles has also pursued deconstruction through artificially replicated or simulated surface degradation. Caroline Evans's *Fashion at the Edge* offers an analysis of the dark side of late twentieth-century fashion and the preoccupation of some avant-garde designers with dereliction, death and decay. Alison Gill, similarly, has examined the work of fashion designer Martin Margiela, who halted garments in a raw state, with seams laid bare and hems unturned, or subjected his designs to wear and tear, introducing rips and frays into new clothing.[32] These garments mimicked the effects of time, and inverted the work of finishing and maintenance, a process similar to that used to fake the effects of age in objects passed off as authentic 'antiques' (in the latter case, the strategy is stripped of its philosophical content).

Like Derrida, Gilles Deleuze provides concepts that can be applied more materially than he may have intended. Deleuze has used the excessive representation of draperies that is characteristic of Baroque painting and sculpture as a framework for philosophical speculation. Folded cloth provides him with a conceptual model for intellectual relationships and correspondences: points that may be distant on a flat cloth can be proximate if that cloth is folded, and the patterns of the fold are endlessly complex, with infinite permutations.[33] The Deleuzian fold has been a fruitful source of inspiration in architectural theory and practice; as Greg Lynn writes, 'as the topological surface folds over and into itself multiple ... points of intersection are possible'.[34] It has been equally provocative within textile practice, where theorists such as Pennina Barnett have explored its implications in a field that is so dominated by surface. If cloth is all surface, then Deleuze's folded draperies provide 'a surfeit of exterior ... a façade freed from its interior', impenetrable proliferations of cloth. (In Chapter 12, Ulrich Lehmann explores Deleuze's ideas further in his analysis of Austrian-Italian designer Carol Christian Poell.)

Depth, im/permeability

Many of the studies discussed thus far have accepted the opposition between superficiality and depth, structure or core, even if they have found the notion of surface worthy of some consideration or investigation. But what happens when, in different ways, the surface/depth binary is broken down, is questioned, or even reversed? We can immediately return to the Deleuzian fold here: if the drapery of the Baroque fold provides an excess of surface, then it is also the case that the fold takes the flat textile, and, by folding it, creates interior spaces: 'topological geometry captures space within its surface'.[35] What is the effect of the *pocket* on the surface/depth binary, as the fabric of a garment, that provides a surface or second skin for the body within, is folded and fixed into a secret and hidden space, one that is often described as 'deep', and that is analogous to the body's own interior spaces? As Ellen Lupton has noted: skin 'flows continuously from the exposed surfaces of the body to its internal cavities'.[36]

If the concept of the fold complicates the surface/depth binary, then what happens when the surface, folded or smooth, is understood as permeable? (See Chapter 13, Alice Barnaby's analysis of light and reflection in the nineteenth-century interior for an interesting application of the notion of surface impermeability.) Permeability suggests susceptibility to breaking down, tearing, or seeping: much of Julia Kristeva's work is concerned with boundaries, and with the transgression of boundaries and the surfaces that form them. In the opening pages of *Powers of Horror*, Kristeva discusses the skin that forms on warmed milk ('thin as a sheet of cigarette paper, pitiful as a nail paring'), describing her gagging, sweating revulsion at this clingy, membranous surface that, because it is neither solid nor liquid, but a confusion between the two, seems to her *abject*, a defilement of categories.[37] But even when surfaces and boundaries are more securely constituted than the skin on milk, Kristeva is concerned with what happens when they dissolve (recalling Mary Douglas's assertion that 'all margins are dangerous' as they are where systems are most vulnerable).[38] Kristeva's category of the 'abject' is the result of such a breakdown: 'it is ... not lack of cleanliness or health that causes abjection but what disturbs identity, system, order. What does not respect borders, positions, rules. The in-between, the ambiguous, the composite.'[39] The ambiguity of the abject, its wrongness, is to do with the breakdown or transgression of boundaries, when within and without migrate and mix – the stuffing coming out of a ripped chair, or blood flowing from torn skin. The breakdown of any surface – frayed garments, chipped paint, flaking varnish – is suggestive of abjection because its role as an impermeable boundary is seen to be under threat of dissolution.[40]

Finally, we should consider what happens when the surface/depth binary is not so much complicated, as reversed. In material practice, surface is usually secondary to form; its processes follow and are dependent upon construction, and an object must exist before it can have a finished surface. Yet Steve Brown, who discusses his work in Chapter 15, forms ceramic sculptures and prints their surfaces at one and the same time – indeed, the surface determines the form. Anthropologist Daniel Miller argues that there is a 'pervasive assumption' in contemporary western culture, emblematized in the term 'superficial', 'which implies that there is a relationship between surface and lack of importance'.[41] He names this tendency 'depth ontology': 'we tend to assume that everything that is important for our sense of being lies in some deep interior and must be long-lasting and solid, as against the dangers of things we regard as ephemeral, shallow or lacking in content'.[42] Miller suggests that not all societies or modes of thought subscribe to the binarism that underpins this view; sometimes surface effects take precedence. In an ethnographic analysis of fashion in Trinidad, tellingly titled 'Style and ontology', Miller documents the conspicuous consumption of spectacular and ephemeral fashion among certain demographic groups, a custom condemned as both superficial and wasteful by some commentators. Yet Miller argues that the practice functions as part of a social and aesthetic system that actively prioritizes the 'superficial' in both objects and interpersonal relations, for good reason. Miller finds a rationale for the behaviour he describes in black diasporic experience:

> The precarious existence given by the condition of slavery precluded any internalization of love, since there was no knowing when this love-object might be wrested from one. The result being a kind of adaptive tendency to keep things on the surface, to refuse any internalization, and thus to minimize one's sense of loss.[43]

In other words, in post-colonial culture, resistance might be expressed by keeping on the surface that which cannot be allowed to become too deep-rooted, in the face of a history of oppression and deracination.[44]

This tendency of the dispossessed to display 'deep' meaning on the surface, or to displace it onto the surface through 'style', is not limited to Miller's Trinidadian example. Dick Hebdige's classic 1979 semiotic account of subcultural style, music and behaviour demonstrates how trivial objects can be elevated to the status of signs, signalling a refusal of norms, and making a claim on alternative identities in the face of poverty and marginalization.[45] In the ripped and torn garb of 1970s punks, crudely decorated with appropriated household objects (safety pins, toilet chains), messages are inscribed upon the surfaces that constitute style (messages that are analogous, as Hebdige suggests more than once, to graffiti). But although these messages are located on the surface, they may carry secret meanings more usually associated with depth: 'our task becomes ... to discern the hidden messages inscribed in code on the glossy surfaces of style, to trace them out as "maps of meaning"'.[46] As Miller notes, the surface/depth metaphor may become 'highly problematic' when 'we encounter a cosmology which may not share these assumptions, and rests upon a very different sense of ontology'.[47] (See Chapter 14, Stephen Knott's investigation of paint-by-number kits, an investigation of surface and identity in a quite different context.)

'In broad cultural terms, there has been a movement away from dialectical relationships, from the oppositions between surface and depth, in favor of an awareness of the oscillating movement from one into the other', wrote Alice Imperiale in 2002 in *Skin*, the collection of writings and images that accompanied the exhibition of the same name at the Cooper Hewitt Museum in New York.[48] This move arguably had its origins in postmodernism (see Chapter 16, Glenn Adamson's exploration of the postmodern surface). In contemporary design digital technologies allow more and more complex articulations of form. Yet surface, whether viewed as continuum or binary, remains a powerful conceptual category through which to investigate material objects, in both the contemporary world and the past. The essays that follow in this volume approach the task from a variety of disciplinary and theoretical positions, but all are unified by a concern for the ways in which surface and finish, however complex their definition, contribute significantly to the characteristics of things.

Notes

1 B. Pell, *The Articulate Surface: Ornament and Technology in Contemporary Architecture* (Basel: Birkhäuser, 2010), Introduction.

2 C. Edwards, *Victorian Furniture: Technology and Design* (Manchester: Manchester University Press, 1993), pp. 32–34; M. Ponsonby, *Stories from Home: English Domestic Interiors*, 1750–1850 (Aldershot: Ashgate, 2007), p. 96.

3 A. Loos, 'Ornament and crime' (1929), in *Ornament and Crime: Selected Essays* (Riverside, California: Ariadne Press, 1998).

4 *Ibid.*, p. 171.

5 Pell, *The Articulate Surface*, p. 8.

6 Le Corbusier, *The Decorative Art of Today* (1925) (Cambridge, Massachusetts: MIT Press, 1987), p. 192.

7 M. Mostafavi and D. Leatherbarrow, *On Weathering: the Life of Buildings in Time* (Cambridge, Massachusetts: MIT Press, 1993), p. 76.

8 M. Wigley, *White Walls, Designer Dresses: The Fashioning of Modern Architecture* (Cambridge, Massachusetts: MIT Press, 1997).

9 Pell, *The Articulate Surface*, p. 7.

10 N. Schor, *Reading in Detail: Aesthetics and the Feminine* (New York: Methuen, 1987).

11 L. Jackson, *Twentieth-century Pattern Design: Textile and Wallpaper Pioneers* (London: Mitchell Beazley, 2002), pp. 45–47.

12 P. Kirkham, *Charles and Ray Eames: Designers of the Twentieth Century* (Cambridge, Massachusetts: MIT Press, 1995). See also J. Anger, *Paul Klee and the Decorative in Modern Art* (Cambridge: Cambridge University Press, 2004); S. Bliss, 'Cubistic claptrap? Erik Magnussen's *The Lights and Shadows of Manhattan* of 1927', in *Silver Studies*, 21 (2006), 113–119.

13 R. Venturi, D. Scott Brown and S. Izenour, *Learning from Las Vegas* (Cambridge, Massachusetts: MIT Press, 1972).

14 M. Wigley, *The Architecture of Deconstruction: Derrida's Haunt* (Cambridge, Massachusetts: MIT Press, 1993).

15 Mostafavi and Leatherbarrow, *On Weathering*, p. 16.

16 M. Douglas, *Purity and Danger: An Analysis of the Concepts of Pollution and Taboo* (1966) (Routledge: London, 1991).

17 *Ibid.*, p. 2.

18 G. Bachelard, *The Poetics of Space* (1958) (Boston, Massachusetts: Beacon Press, 1994).

19 *Ibid.*, p. 68. Bachelard was writing in the late 1950s, and his analysis reflects the gendered attitude to domesticity of his day.

20 *Ibid.*, pp. 101–102. For another interesting phenomenological approach, see J. Pallasmaa, *The Eyes of the Skin* (London: Academy Editions, 1996).

21 S. Walker, *Sustainable by Design: Explorations in Theory and Practice* (London and Sterling, Virginia: Earthscan, 2006), p. 87. See also S. Walker, 'The environment, product aesthetics and surface', *Design Issues*, 11:3 (1995), 15–27.

22 Walker, *Sustainable by Design*, p. 87.

23 G. McCracken, *Culture and Consumption: New Approaches to the Symbolic Character of Consumer Goods and Activities* (Bloomington and Indianapolis: Indiana University Press, 1988), pp. 31–43.

24 O. Wilde, *The Picture of Dorian Gray* (1891) (Harmondsworth: Penguin Books, 1985), p. 45.

25 C. Pajaczkowska, 'On stuff and nonsense: the complexity of cloth', *Textile: The Journal of Cloth and Culture*, 3:3 (2005), 242. See also J. Entwistle, *The Fashioned Body: Fashion, Dress and Modern Social Theory* (Cambridge: Polity, 2000); J. Entwistle and E. Wilson (eds), *Body Dressing* (Oxford and New York: Berg, 2001).

26 P. Barnett, 'Folds, fragments, surfaces: towards a poetics of cloth', in *Textures of Memory: The Poetics of Cloth* (Nottingham: Angel Row Gallery, 1999), p. 28.

27 G. Vigarello, *Concepts of Cleanliness: Changing Attitudes in France Since the Middle Ages* (1985) (Cambridge: Cambridge University Press, 1988), pp. 58–71.

28 K. Hauser, 'The fingerprint of the second skin', in C. Breward and C. Evans (eds), *Fashion and Modernity* (Oxford: Berg, 2005), pp. 153–170.

29 B. Elliot and J. Helland, *Women Designers and the Decorative Arts 1880–1935: The Gender of Ornament* (Aldershot: Ashgate, 2002), p. 11.

30 T. Gronberg, 'Decoration: Modernism's "other" ', *Art History*, 15:4 (1993), 547–552.

31 V. Kelley, 'The interpretation of surface: boundaries, systems and their transgression in clothing and domestic textiles, c.1880–1939', *Textile: The Journal of Cloth and Culture*, 7:2 (2009), 224–225.

32 C. Evans, *Fashion at the Edge: Spectacle, Modernity and Deathliness* (London: Yale University Press, 2003); A. Gill, 'Deconstruction fashion: the making of unfinished, decomposing and re-assembled clothes', *Fashion Theory*, 2:1 (1998), 24–29.

33 Barnett, 'Folds, fragments, surfaces'; A. Gavin, 'The matter with de Clérambault: on the baroque obscene', *Textile: The Journal of Cloth and Culture*, 7:1 (2009), 56–67.

34 G. Lynn, *Folding in Architecture* (1993) (Chichester: Wiley-Academy, 2004), p. 28.

35 *Ibid.*, p. 28.

36 E. Lupton, 'Skin: new design organics', in E. Lupton (ed.), *Skin: Surface, Substance and Design* (New York: Princeton Architectural Press, 2002), p. 29.

37 J. Kristeva, *Powers of Horror: An Essay on Abjection* (1980) (New York: Columbia University Press, 1982), pp. 2–3.

38 Douglas, *Purity and Danger*, p. 122.

39 Kristeva, *Powers of Horror*, p. 4.

40 Kelley, 'Interpretation of surface', 228.

41 D. Miller, 'Style and ontology', in J. Friedman (ed.), *Consumption and Identity* (London: Harwood, 1994), p. 71.

42 *Ibid.*, p. 71.

43 *Ibid.*, p. 85.

44 For further exploration of surface display as a strategy of the dispossessed in a post-

colonial context see Thomas Leddy's discussion of the Chicano *rasquache* aesthetic: T. Leddy, 'Sparkle and shine', *British Journal of Aesthetics*, 37:3 (1997), 259–273.

45 D. Hebdige, *Subculture: The Meaning of Style* (London: Routledge, 1979).

46 *Ibid.*, p. 18.

47 Miller, 'Style and ontology', p. 71.

48 A. Imperiale, 'Digital skins: the architecture of surface', in Lupton, *Skin*, p. 55.

2 ✧ Surface as practice

Lesley Millar

Touch cannot be in opposition to itself, can never be perceivedas surface or source, but an acknowledgement that actual is mutual – a conjoining of two.[1]

WHITE, soft, newly fallen snow, dotted with footprints revealing the semi-translucent ice beneath the surface – this is the memory Japanese graphic designer Kenya Hara wished to bring to mind for all those who touched the cover of the programme he designed for the Nagano Winter Olympic Games. The paper was white and puffy and allowed the characters and letters to be debossed, a process by which a heated press is applied and the fibres of the concave, debossed area are partially melted, becoming translucent – like ice.[2] As we touch the surface of the paper we reconnect to the hidden realm of our sensory perception. We engage with the world through touch rather than merely encounter it in terms of vision alone (Plate 1).

In our most intimate relations we acknowledge and engage in this non-verbal form of communication. It is one of the ways, perhaps the most effective, that we can come to fully know another: 'Kisses tattoo my skin, / leaving a Braille only your fingertips read.'[3] In reading that Braille we are accessing our own personal vocabulary of touch, a vocabulary that we have been accumulating from the moment we were born. Without such points of tactile reference there would be no textural reading, understanding, or dialogue.

As babies we learn to define and refine our relationship with our surroundings through licking, touching, smelling, hearing and seeing, and throughout our lives we continue to experience the world through the acknowledged or unacknowledged vocabulary of our senses, in particular our sense of touch. Our skin is the active medium through which we process information; the surfaces of our tongue, eardrum and eye allow us to taste, hear and see. Yet once we have acquired a verbal language, we institute a hierarchy of the senses, with touch, taste and smell regarded as the lower senses.[4] We rarely acknowledge how much we understand through our haptic awareness; there is an intimacy, a privacy surrounding our sensory experiences, their very bodily nature a potential source of embarrassment. Yet, 'a human being is a bundle of very delicate receptor organs and at the same time an image-generating organ equipped with a vigorous memory-playback system. An image generated in the human brain is a spectacle orchestrated through multiple sensory stimuli and revived images.'[5]

It is this 'playback system' that I wish to explore. As a maker I have a life-time's internal archive of touching to know and to understand which I can, when necessary, access and translate through an informed imagination. Antony Gormley writes of 'the intuition of the un-thought known, the feel of a handle, the way in which a finger explores and finds the place for its crookedness in the ring of a cup handle'.[6] These are physical memories that can guide us when conscious memory fails us. According to Rilke: 'Hands have stories, they even have their own culture and their own particular beauty. We grant them the right to have their own development, their own wishes, feelings, moods and occupations.'[7] If we look at our hands, we see how our lives are traced and embedded in the lines, lines that have been created by our habitual use, which have become 'the embodiment of information on the flesh itself' (Plate 2).[8] The memory of what we do is captured forever in what we are.

Touching the surface of an object, feeling its weight in the hand, exploring its texture with the fingers is a profoundly sensual experience. The love of textural engagement is present everywhere in traditional and contemporary Japanese life. The Japanese artist Teruyoshi Yoshida describes how:

> We take each dish in our hands and either consciously or unconsciously, take pleasure in feeling its texture. We admire the colour and shape while placing the dish inside our palms, and we bring it to our lips to feel with further delicacy. To see by feeling, to feel by seeing – our tactile perception includes a visual aspect in touch, and this form of perception is what we are accustomed to as a way of perceiving beauty.[9]

This visual tactility includes the term '*tanagokoro-de-itsukushimu*' (to cherish the warmth of the palms) which is used by Japanese when appreciating a work of art. Although this is not a quality we in the West consciously look for in a work of art, instinctively we respond to it. In a review in the *Independent* newspaper, Frank Auerbach is quoted as saying about his work: 'I felt there was an area of experience – the haptic, the tangible, what you feel when you touch somebody next to you in the dark – that hadn't perhaps been recorded before.' To which critic Tom Lubbock responded: 'Auerbach's painting is to be looked at, but what it depicts is predominantly not visual … you look at the paint, and see the pressure, strain etc. that his hand has imbued it with. You feel that. The picture's gestures are transmitted to your own body. That's where these Auerbach's happen.'[10] In other words, you *look* at the work, and your body *feels* it, a sensation described by Grayson Perry as a 'visceral bodily response'.[11]

Touch is immediate, and as such can belie the depth of communication it offers. When we touch an object that connects us to history, our history, the history of society and the history of the object, we are touching the past. Yet in order to access these haptic insights, we have to learn the language of our fingers. Edmund de Waal described this physical connection as:

> The bodily memory of picking up a Korean tea bowl from the twelfth century and feeling it and the bodily memory of other things in different cultures and different places. The simple somatic thing about feeling an object and feeling the memory of the making of that object can be overwhelming.[12]

When we run our fingers over a sacred object, or a historic vessel, there is an exchange of energies. We are experiencing a surface that other hands have experienced and changed through touch. We are touching their touch: to touch is also to be touched.[13]

We also contact the maker, whose touch is imprinted within the work. Textile artist Maxine Bristow describes her constant handling of the cloth as she works:

> The laborious working of row upon row of stitching, the hand turning of buttonholes and cracking of gesso encrusted cloth, every centimetre of the surface within my work, bears the trace of my own DNA trapped within the fibres of the cloth.[14]

As we handle the work, our touch and the touch of the maker conjoin and become one, no matter how ancient the object and its surface (Plate 3).

Locard's Exchange Principle, the basis of modern day forensic science, has identified that every contact leaves a trace. It is in our relationship with textiles, the everyday physical engagement between our body and cloth, that the intimacy and potency of the mutuality of touch is most clearly articulated. Through the complicity of the senses it is as if we can already feel the cloth before we have touched it. This sensory approach is the unspoken narrative of textiles, and constitutes a centrifugal shift from seeing, which is intellectual, to touching, which is experiential.The combining of textiles with new technology to create e-textiles and bio-responsive cloth – 'intelligent' textiles – has moved the 'site of engagement' from the hand to the body itself.[15]

If you talk about textiles to someone who works with cloth, before too long they will begin to describe with their fingers. I pick up a yarn and rub it between my fingers, I pull it and stretch it and I know if it is appropriate for the task in hand. I can already see the end product in my mind from the touch. The textile designer Reiko Sudo has described how she approaches the creative process: 'The first image that comes to mind is the feel and touch of the material, its texture. Before considering its use, I always begin with how coarse or smooth it feels. I use my fingertips.'[16] Note that she says the first *image* in her mind is one of touch and texture.

Examining the relationship between touching, making and knowing, and how the intertwined processes can be described and understood, brings a change of emphasis, away from a hierarchy of communication which prioritizes the verbal, and then the visual. Inherent in the act of making is a body of knowledge formed from the intuitive decisions involved in the process. These are decisions that are not verbally predetermined but arrived at in collaboration with the material itself. Japanese textile artist Yuka Kawai describes her working process thus: 'In weaving the cloth I do not so much form the shape according to my will but rather I value the shape which arises from the weaving, from my hands. I believe in the tacit knowledge of our hands.'[17]

The difficulty at the beginning of the twenty-first century arises in maintaining the currency of the transferability of that 'knowledge gained in the hand through touch and movement'.[18] This process was achieved in the past through the relationship between mother and daughter, teacher and pupil, master and apprentice. This relationship was built up over time, years of watching, handling and making, slowly acquiring intuitive skills and material understanding. What resulted was the creation of a memory bank that operates as a discriminator in terms of our intuitive responses. As we develop tactile experience we are consciously accruing experiential knowledge which can act as a determinant in our appreciation of aesthetic/tactile experience (Plate 4). I am lucky to have had the opportunity to build this personal memory bank; however, what concerns me is the systematic exclusion of the acknowledgement of touch from our lives, the loss of opportunities to know about a thing through touch, and the consequences of such an absence. Our need to touch has been acutely highlighted by the success of the touch screen. Everywhere there are fingers delicately touching, almost caressing membranes through which they can access information. But it remains the case that, by touching a digital image on the screen, we do not touch the object or person, but rather the screen. The more we attempt to control our environment and our interaction with the physical world, and deny the fundamental importance of textural experience, the more we risk losing communication achieved through attention to the senses. Schools in the UK barely include craft processes within the curriculum, and craft courses at art institutions are disappearing. Anthropologist Daniel

Miller has observed: 'What we now attempt to inculcate in children through explicit pedagogic teaching, based largely in language, had previously been inculcated largely through material culture.'[19] The result is that as a society we have less tactile knowledge than ever before and we are no longer used to learning material awareness through the touch of the hand, what Richard Sennett describes as 'the conversion of information and practices into tacit knowledge'.[20]

This applies not only to those of us working in visual arts and crafts but to everyone, especially to those whose work involves 'surface as practice'. A surgeon described to me how his job involved tactile experiences: the appreciation of what a joint *feels* like is very informative in terms of diagnosis and future treatment. At school in the 1960s he had, as a matter of course, learned through touch and through making, in pottery, metal work and wood work. Through these experiences he began the translation of the 'feel' of something into understanding and knowledge. When questioned he acknowledged that his haptic sensibilities are taken for granted. His appraisal of textures and surface as a means of identifying shapes without visual help is a function that lends itself to critical analysis.

As long ago as 1921 the founder of Futurism, Filippo Tommaso Marinetti, identified the importance of informed touch in his Manifesto of Tactilism, creating a scale of six levels of touch:

> **First category: extremely confident touch, abstract, cold.**
> Sandpaper, Silver-coated paper.
> **Second category: touch without heat, persuasive, reasoning.**
> Smooth silk, Silk crêpe.
> **Third category: exciting, lukewarm, nostalgic.**
> Velvet, Wool from the Pyrenees, Wool, Silk-wool crêpe.
> **Fourth category: almost irritating, hot, determined.**
> Granulous silk, Plaited silk, Spongy cloth. Second scale, volumes.
> **Fifth category: soft, hot, human.**
> Suede, Horsehair or dog hair, Human hair, Marabou.
> **Sixth category: hot, sensual, spirited, affectionate.**
> Rough iron, Soft brush, Sponge, Wire brush, Plush, Human or peach fuzz, Bird down.[21]

From these categories he created boards with surfaces constructed from arrangements of materials of different tactile values, for what he designated as 'hand journeys'. The hand wandered over the surfaces, following trails and experiencing a succession of suggestive sensations whose rhythm, in turn languid, cadenced, or tumultuous, was regulated by exact directions. In this way Marinetti transformed a collage of materials into a montage of tactile experience.

Marinetti's experiments and ideas foreshadowed the recent work of Japanese designer Masayo Ave (Plate 5). Ave has spent several years in different countries, exploring the articulation of texture, the Sound of Materials.[22] The Japanese language is rich in its use of onomatopoeic description, for example *fuwa fuwa* is a soft fluffiness, fluffy clouds. Ave began this project after realizing that other languages were not so rich in onomatopoeia, and she has been working with groups of children and adults, identifying and describing textures through mutually agreed, invented onomatopoeic descriptors, creating a haptic dictionary. An example of the newly created haptic words is *hara*: velvet which is smooth and sensuous when rubbed in one direction and rough when rubbed in the other. And from these sounds she has gone on to design and develop objects and surfaces that, in turn, illustrate the sounds, thus completing the circle of communication. Her work explores the mutuality of touching and being touched, the bodily response translated into an emotion, a feeling, or the possibility of the transmission of information through the contact between one surface and another (Plate 6).

It is important to stress that I am not arguing for an artificial reconstruction of past practice, which would be as impossible as it is undesirable. What I suggest is parity between an approach based on the primacy of verbal language and one based on sensory language (Plate 7). What is required is a plurality of communication, achieved equally through access to textural intelligence obtained through the manipulation and touching of materials, and through screen based and virtual experience. An example of this plurality is the German scientist who noticed that water always ran off the surface of the lotus leaf. Investigation revealed tiny bumps on the surface preventing the water sticking. This led, through nano technology, to an external house paint with a self cleaning surface. Thin layers of titanium dioxide, when charged with sunlight, break down dirt into particles and create sheets of water when it rains. These sheets roll off leaving a clean, dry surface. All this was enabled through a man's fascination with surface as practice.

It is argued that new technology enables us to 'challenge our role as a passive audience and actively participate in a non-verbal conversation … we have the opportunity to let our bodies become a resonating organ of our surroundings … we have the tools to establish a visceral connection to our mind'.[23] Indeed we do have the tools, and have done since the emergence of the first sentient beings; we can reach out and allow our fingers to do all this, and so much more.

Notes

1 H. Chadwick, 'Lumina Delights', in *Enfleshings* (New York: Aperture Foundation, 1989), p. 69.

2 K. Hara, *Haptic* (Tokyo: Asahi Shimbun, 2004), p. 28.

3 A. Stubbersfield, 'Six colours: no.4 midsummer', in *Joking Apart* (Gwent: the Collective Press, 2006), p. 44.

4 J. Ree, *I See a Voice* (London: Harper Collins, 1999), p. 351.

5 K. Hara, *Designing Design* (Baden: Lars Müller Publishers, 2007), p. 156.

6 A. Gormley, 'Thinking about Naoto Fukasawa', in N. Fukasawa (ed.), *Naoto Fukasawa*(London: Phaidon, 2007), p. 118.

7 R. M. Rilke, *Auguste Rodin* (New York: Archipelago Books, 2004), p. 44, as quoted in J. PallasmaaJuhani, *The Thinking Hand* (Chichester: John Wiley and Sons, 2009), pp. 28–29.

8 P. Lichty, 'Building a culture of ubiquity', *Intelligent Agents*, 8:1 (2008), www.iamwhateveryouwantmetobe.com, accessed 31 January 2010.

9 Quoted in L. Millar, *Through the Surface* (Farnham: University College for the Creative Arts, 2004), p. 95.

10 T. Lubbock, 'In the thick of it', *Independent*, 9 November 2009.

11 G. Perry, speaking at 'Amateurism, Art and Absorption', talk at London Literature Conference, 2008, reported in *Crafts*, 214 (2008).

12 E. de Waal, 'Conversations on Making', talk at Siobhan Davies Studios, 2009, www.siobhandavies.com/conversations/, accessed 30 November 2009.

13 P. Rodaway, *Sensuous Geographies: Body, Sense and Place* (London: Routledge, 1994), p. 41.

14 M. Bristow, *Continuity of Touch: Textiles as Silent Witness*, www.cpara.co.uk/events/repeatrepeat/labour/Bristow.html, accessed 24 October 2009.

15 Lichty, 'Building a culture of ubiquity'.

16 Hara, *Haptic*, p. 182.

17 Quoted in L. Millar, *Cultex* (Oslo: Kunstcentralen, 2009), p. 60.

18 R. Sennett, *The Craftsman* (London: Allen Lane, 2008), p. 35.

19 D. Miller (ed.), *Materiality* (Durham, North Carolina and London: Duke University Press. 2005), p. 6.

20 Sennett, *The Craftsman*, p. 50.

21 F. T. Marinetti, read at the Theatre de l'Oeuvre (Paris) and the World Exposition of Modern Art (Geneva), published in *Comoedia*, January 1921.

22 M. Ave, 'Haptic', lecture at RIBA, London, 7 May 2008.

23 A. M. Skriver-Hansen, 'The body as interface', *Intelligent Agents*, 8:1 (2008), www.iamwhateveryouwantmetobe.com, accessed 31 January 2010.

PART I

TIME, WORK

3 ✧ Layering: materiality, time and touch in Japanese lacquer

Christine M. E. Guth

Layering has long functioned as a tacit mode of socio-cultural and political communication in Japan. Nowhere has this been expressed more literally than in the so-called 'twelve layer dress' or *jûnihitoe*, which could in fact comprise as many as twenty silk robes worn over one another. The *jûnihitoe* was a mark of aristocratic status from around the tenth century, owing to the sheer quantity of fabric and the amount of labour required to produce it. Adding to its cultural significance were the subtle differences in colour that distinguished each layer, the dyeing and selection of which served as a means for court women to express their individuality. Social and economic changes led to considerable simplification of dress in the sixteenth century when the small-sleeved undergarment, or *kosode*, was adopted as the principal outer garment among both men and women of all classes. In spite of this, layering continued to serve to map social distinctions, albeit in more subtle and sometimes subversive ways. When the hundreds of yards of silk fabric used for the *jûnihitoe* were eliminated, something of the same effect was achieved in the *kosode* by a costly and time-consuming elaboration of surface. This display of material and technical virtuosity involved the application of multiple, overlapping decorative techniques including tie-dyeing or *shibori* (*kanoko* and *nuishime*), hand-painting with subtle shading effects, the use of gold leaf (*surihaku*), and embroidery.[1] Like the multiple layers of the *jûnihitoe*, this style of treating the *kosode* had both visual and tactile appeal.

A third form of sartorial layering inverted the principles just described by building meaning and value into the undergarment, *juban*, or the lining of the short jacket, *haori*, that were fashionable in the eighteenth century. Practised primarily by men of merchant class background, this strategic subterfuge allowed them to evade the sumptuary laws that prohibited them from wearing silk. One surviving nineteenth-century *juban* is entirely embroidered in cotton twill of multiple colours so as to form a panoramic view of the city of Edo.[2] While made of authorized material, the cost of this embroidery would most likely have outstripped that of a comparable undecorated silk garment. The woodblock print of c. 1795 by Utamaro illustrated here shows a courtesan holding a *haori* open for a departing client (Plate 8). The garment's exterior is unremarkable, but its decorative lining features an irreverent ink painting of the Zen founder Daruma on silk, signed by Suzuki Rinsho, a noted artist at the time. This embellished interior would only have been displayed when its wearer chose to disrobe in the presence of friends and intimates. The communicative power of this fashionable ruse followed from a dynamic of revealing and concealing wealth. Complicity in a shared semiotic code fostered a sense of merchant class solidarity, but also the development of creative individuality within that class.

As the foregoing suggests, the expressive and aesthetic dimensions of layering have been examined in the context of Japanese dress but these investigations have not extended to consideration of the roles that the phenomenon played in other media or at other times.[3] This essay proposes that layering was part of a larger discursive system in early modern Japan, one especially significant in the consumption of Japanese lacquer. Bourdieu has famously argued that distinction is created and maintained through the manipulation of the sign, rather than the functional value of material goods.[4] Through a series of close readings of Makie, Negoro and Ritsuô, three styles of lacquer popular between the sixteenth and nineteenth centuries, I hope to draw attention to the ways that layering materialized power relations within early modern society.

Lacquer is a natural plastic (urushiol) extracted from the lac or *rhus verniflua* tree that polymerizes in a warm humid environment, rendering objects covered with it hard and impervious to moisture. It was used to produce many of the articles deemed essential to the lives of the aristocratic and warrior elite. Food vessels, furnishings for the domestic interior, decorative architectural panels, and even saddles and stirrups were fashioned from wood to which multiple coatings of lacquer were applied.[5] By the end of the sixteenth century, lacquer use had grown such that local trees could no longer supply demand, and raw sap had to be imported from South-East Asia. Consequently, the raw sap circulated as a specialized currency within the monetized economy of early modern Japan. In 1567, for instance, Oda Nobunaga's future son-in-law, as a sign of his wealth, presented the warlord with 1,000 buckets of lacquer.[6] Recent scientific analysis has revealed that some lacquer makers used imported lac for the base strata while reserving locally sourced lac, which was deemed to be of finer quality, for the final coatings.[7]

The profligate use of lacquer was key to the cultural meaning of the intricately decorated dowry sets required for marriage between members of the warrior elite throughout the Tokugawa period (1615–1868).[8] These might include more than fifty pieces – palanquins, furniture, utensils for food and drink, cosmetic cases, writing implements, and games among them – all featuring a common design motif rendered in sprinkled gold and other materials. The acquisition and display of this profusion of shimmering lacquer dowry goods, visibly marked with one or both family crests, signalled their owners' wealth and rank. We might imagine that the seemingly alchemical transformation of liquid lacquer into a permanently hard, glossy surface invested the dowry articles made from it with auspicious properties emblematic of the enduring, integrative experience of marriage.

The quantity of lacquer and gold used in the creation of dowry goods was key to their status as signifiers of luxury and power. Although lacquer was applied to many materials, wood was the most common support for domestic wares. The liquid lacquer was applied in layers, some of them mixed with other substances to build up and colour the body in various ways.[9] Each application of liquid lacquer was followed by curing in a dust-free humid environment before the next layer was applied. Polishing was carried out between each coat to ensure evenness. In the final stages, the outlines of a design were transferred from paper, usually with a brush, and the contours filled in by sprinkling gold particles of various sizes and colours with a bamboo tube. This produced what came to be known as Makie, 'sprinkled picture lacquer'.

Makie lacquers were often made more sumptuous still by the elaboration of the pictorial motifs on the surface with metals, mother-of-pearl and other precious materials. Such elaboration altered the otherwise smooth polished surface of the finished lacquer in such a way that it actually rendered some articles unfit for use. This tension between form and function is illustrated in the table illustrated here, whose raised decoration made it unfit for writing, although it might still serve as a showy support for a book or scroll (Plate 9). Like this elegant table, despite their apparently utilitarian purposes, dowry sets handed down among members of the Tokugawa ruling elite exhibit a remarkable state of preservation, suggesting that they too were intended chiefly for display and as family heirlooms rather than for household use.

Functionally speaking, covering a wooden article with so many coats of lacquer and other precious substances is unnecessary, so it may be argued that this production technique was itself part of a discursive system that valorized superfluity. In other words, a calculated, excessive use of costly materials was built into Makie's production and consumption. Although consumers might not see each stratum of a finished box or bowl, they could feel the thickness and varied surface textures of the vessel and see the flakes of gold and other metals embedded within it through the transparent lacquer skin. They would also be aware of the way that Makie materialized time and craftsmanship. Completion of the equipage of ostentatious pieces commissioned from Kôami Nagashige for the wedding of Chiyohime, the daughter of the third Tokugawa shogun, for instance, required three years, from 1637–39.[10] So important was lacquer in the lives of the warrior elite that most feudal lords retained lacquerers to supply their needs. The Kôami family, generations of whose members served the Tokugawa shoguns, filled no less than eight major wedding commissions between 1620 and 1649 alone.[11] The astounding cost of such sets can be imagined from the diary of Kôami Iyo who describes in detail the process of making a single lacquer writing box in 1718. The total for labour and materials came to 53 *ryô*, a sum equivalent to enough rice for 73 individuals each to live for a year.[12]

Possession of materials associated with status was highly regulated in the Tokugawa period. Although warriors were at the top and merchants at the bottom of the official social hierarchy, growing merchant affluence blurred these distinctions, and the government repeatedly issued sumptuary laws in order to discourage extravagance among officials and prevent merchants from translating money into cultural capital. Regulations regarding the materials and dyeing, embroidery and weaving techniques used in making clothing were issued repeatedly. And when subjects circumvented these by wearing expensive undergarments or dyed silk linings, the government issued more explicit directives.[13] Lacquer of all kinds, but especially Makie, was also a target of such hierarchical regulation, with the amounts of gold that could be used by members of the warrior elite determined by rank.[14] Further regulations banned townspeople from using intricate gold designs on their furnishings or saddles, and even discouraged the sale of lacquered chopsticks in rural villages.[15]

In contrast to Makie, which assumed value in relation to its excessive materiality, Negoro style lacquer accumulated meaning through the shedding of layers. Lacquer conceals a wooden substrate that, without its protective envelope, deteriorates with time, and it was the revelation of this vulnerability that gave Negoro ware its cachet. The process of dematerialization followed from the fact that utilitarian, undecorated matte red wares were made by applying to a wooden core a base coat of black lacquer with only thin coats of lacquer coloured with iron oxide or cinnabar red over it. Because red was an auspicious colour, lacquered trays, boxes and serving utensils made in this way were widely used in Buddhist temples and Shinto shrines. With time and cumulative use, however, the thin red coating wore away to reveal the dark coat beneath. Age and wear traced changes on the surface of the lacquer in random and unpredictable ways that created accidental juxtapositions of red topcoat and black undercoat visible to the eye, though barely felt (Plate 10). These transformations gave Negoro an affective quality that was not the mark of the artistic hand but rather expressive of individual life-cycle. At the same time, by compromising their commodity value, the damage to the surface also provided a means whereby these functional utensils became intrinsically sacred.

The interior of a lacquer container could also hold the memory of articles left within it over a long period of time. A twelfth-century red lacquer tray from Nara's Kasuga Shrine, now in the Metropolitan Museum of Art, for instance, still retains the tangible traces of its history in the form of the faint traces of grime built up between the contours of the Buddhist scrolls it held for centuries.[16] The relationship between these signs of wear and the value of Negoro is by no means obvious, however. Indeed, physical degradation might lead to quite the opposite conclusion – that is, that the worn object should be repaired or even discarded. As the essayist Tanizaki Jun'ichirô, writing of the Japanese taste for such artefacts in his 1933 essay *In Praise of Shadows*, acutely observed, the '"sheen of antiquity" was nothing more than the glow of grime'.[17] To fix the meaning of surface abrasion or dirt in a positive way depends on cultural negotiations that become common knowledge. A passage in Yoshida Kenkô's *Essays in Idleness*, written in the 1330s, about the decision not to alter an unsightly chest that had been handed down from ancient times, because it had 'by its very dilapidation become a model' suggests that there was long-standing recognition in Japan of the social significance of wear, because of the implicit relationship between material deterioration and pedigree.[18]

Although Kenkô may not have been writing specifically about worn lacquer, the admiration that Negoro commanded among later collectors was similarly related to what might be called 'object memory'. Time-worn lacquer gave newly affluent merchants access to the cachet of elite heirlooms. The patina, moreover, made it possible for them to feel the lacquer's age without actually having experienced the external physical world of which it was a part. Its value was all the greater when it was narrativized by identifying the individual owner or temple from which it had come: the lacquer tray in the Metropolitan Museum of Art, for instance, would command far less admiration without its distinguished provenance from Nara's historic Kasuga Shrine.

Although wooden articles made with a black undercoat and red overcoat of lacquer have a long history of use in Japan, the label Negoro was not applied to them until the seventeenth century.[19] Exactly how this came about is a subject of scholarly dispute, but the importance in Japan attributed to creating 'biographies' of things surely played a role.[20] This also may help to explain why, according to one etymology, a style of lacquer that so clearly betrays signs of physical degradation came to be named after a temple that was cruelly destroyed in the sixteenth century by the warlord Hideyoshi. Such linguistic accounts are important in bridging the gap between words and things in such a way that they tell a larger human story. As Stephen Greenblatt has observed in an essay on resonance and wonder, 'wounded artefacts may be compelling not only as witnesses to the violence of history, but as signs of use, marks of the human touch, and hence links with the openness to touch that was a condition of their creation'.[21]

It is difficult to determine exactly when the fashion for Negoro wares first developed, but situating the value of wear and dirt of this style of lacquer in its own past was no doubt mediated by the influential writings of merchant class tea ceremony masters whose thinking was deeply informed by Buddhist belief.[22] In the late fifteenth century tea practitioner Murata Shukô helped to carry out this socio-religious aestheticization of the process of damage and decay through his writings on the relational nature of perfection. How can we understand perfection without seeing it in contrast to that which is imperfect? This question underlies one of his best-known aphorisms, 'The moon is not pleasing unless partially obscured by a cloud'.[23] Such statements both shaped and reflected the use and display of cracked or mended ceramics alongside refined and ornate Chinese porcelains in the tea ceremony. For Buddhists, the material world was understood to be transient and insubstantial, and articles such as worn lacquer served to point to this truth by reminding their owners of their own mortality every time they handled them.

The taste for Negoro also may have been mediated by the prestige ornate Chinese carved lacquer wares had assumed among Japan's warrior elite since the fifteenth century.[24] Chinese carved lacquer self-consciously mobilized the material and aesthetic qualities of layering to create showy, richly textured, polychromatic articles ranging from trays to boxes of various shapes and sizes. Each one was made by repeated applications of lacquer, sometimes impregnated with different colours, usually red and black, sometimes with an additional stratum of yellow. This built up a very thick body that was then carved in complex curvilinear decorative patterns so as to reveal the coloured bands beneath the surface.[25] While these carved lacquers are more weighty and ornate in form and décor than Negoro, they share a predilection for alternating layers of black and red.

The ability to read the material and symbolic code that made physical degradation culturally valuable, which required seeing beyond external appearances to the intrinsic qualities of things, was not within everyone's grasp, however. The prestige of patina, to quote Grant McCracken, was 'part of a larger process in which society turned money into status ... a hidden code immediately intelligible to those of genuine standing and well concealed from all but the most sophisticated pretenders'.[26] Similarly, in early modern Japan, appreciation of Negoro began as part of a strategy through which merchant class practitioners of the tea ceremony legitimated their consumption practices while simultaneously gaining the moral high ground over the hereditary feudal elite.

The tea ceremony in which the value of imperfection first developed was initially a subculture built around exclusivity and being 'in the know', but over time, enough people bought into its sign system that it became a powerful design force, both in the re-evaluation of articles handed down from the past and in the production of new ones. Indeed, Negoro ware became so sought after that by the nineteenth century, lacquer makers were deliberately subjecting their creations to an artificial process of wear that radically shortened the time required to acquire the mottled individuality that made them such desirable commodities. A case in point is a tray acquired in 1882 by the British collector John Quin, now in the Royal Botanic Gardens at Kew, that simulates the very qualities that made Negoro unique.[27] Here marks of wear create a self-consciously decorative pattern of alternating red and black.

For a third example of the aesthetics of layering in Japanese lacquer, I turn to the work of Ogawa Haritsu (1663–1747), also known as Ritsuô, who lent his name to a style that was much imitated over the course of the eighteenth and nineteenth centuries.[28] In keeping with the growing spirit of scientific enquiry of the times, Ritsuô sought to demonstrate through the medium of lacquer his understanding of animals, minerals and plants and the effects of time on them through exacting attention to nuances of form, surface texture and colour. Ritsuô style lacquer wares are distinguished by their simulation of the look and feel of materials including ceramics, bamboo, metal, inksticks and woodgrain. His creations testify to an awareness that knowledge formation requires getting beneath the surface of things, understanding them both inside and out, through touch.

While document and writing boxes figure prominently among Ritsuô style wares, these also include *inrô*, small containers for personal possessions suspended at the waist that became fashionable accessories for affluent men of all classes beginning in the seventeenth century.[29] By the eighteenth and nineteenth centuries, this was probably the form in which lacquer was most widely experienced in daily life, albeit solely by men. Because of their small size, *inrô* were less important as functional containers than as markers of individual taste. As such they articulated a new site of circulation for lacquer involving a slippage from the world of family heirloom or ceremonial vessel to that of passing fashion, or *fûryû*. That they were conversation pieces and collectors' items is clear from the feudal lord Matsuura Seizan's journal of 1821–41. Well known for his sartorial extravagance, he described how people would follow him to see what he was wearing at the waist. 'Naturally,' he wrote, 'it wouldn't have done for me to have worn the same things under these circumstances … [so] I was continually changing my accessories.'[30]

Inrô offered a very different physical and tactile experience of lacquer from Makie or Negoro wares. Because they were intended to hold small personal items such as medicine or seals, they were kept close to the body and handled often; Seizan's diary reveals the pleasure he derived from fondling these sensuous, diminutive objects.[31] Further pleasure followed from the fact that many *inrô* playfully simulated the forms and textures of familiar items. A Ritsuô style *inrô* now in the Victoria and Albert Museum, for instance, takes the form of a miniature temple gong (Plate 11).[32] It convincingly reproduces the appearance of deeply eroded metal, with the inscription around its circumference rendered illegible and the upper edge partially worn away. In simulating the effects of wear, its creator self-consciously sought to demonstrate his command over nature itself, bypassing the functionality normally required to produce an antique. The process of time, like the object itself, is compressed. This teasing deception was part of its appeal. Ironically, the apparent temporal compression in the *inrô* added to the time required to read it, taking in and connecting the various layers of meaning.

Unlike artificially aged Negoro vessels, Ritsuô wares involved a knowing collusion between artist and consumer, a suspension of disbelief that brought pleasure in the consumption of a fashionable commodity. By creating a simulacrum that cannot be used as intended, something that is, moreover, permanently caught in the process of destruction, Ritsuô manipulated lacquer both materially and conceptually to make people think about the process of time, but in an ironic way, since the article itself embodied a passing fashion. The temple gong made miniature evokes the *idea* of a gong just as its surface abrasion evokes the idea rather than the reality of erosion.

Ritsuô's *inrô* mobilizes layering physically as well as metaphorically, through the device of *mitate*, which, in this context, might be described as a technique that creates surface tensions by using one thing to say something else.[33] Common to both visual and material culture in the early modern period, *mitate* was a form of glossing that often served to validate mundane objects by superimposing on them meanings drawn from more elevated contexts such as classical literature or Chinese culture. The secular *inrô*'s silent referencing of a sacred temple gong was thus part of a larger system of signification that wearer and viewer alike would have recognized.

The wealth and variety of lacquered objects in the early modern period testify to a rapidly growing economy in which the specialized consumption habits of feudal lords and affluent merchants figured significantly. In this chapter I have proposed that layering may serve as a matrix for understanding the way lacquer wares functioned as sites for constituting and communicating the socio-cultural, political and economic identities of their makers and users. Because of its varied modes of application, colouration and decoration, lacquer had the capacity to emphasize different physical and metaphysical aspects of the articles it covered. The multiple of meanings encoded in its layers should be understood in a nuanced relation to one another, as part of a larger discourse on materiality, time and touch. The strata built up to give a Makie ware its distinctive artefactuality contribute to its value by rendering it durable and demonstrably opulent. But if power is required to mobilize the forces of time, skill and money to accomplish this, such power also may be suppressed through erasure, especially when moral value is imputed to such activity. This subversion was promoted by merchants who, defining good taste as a denial of opulence, embraced Negoro wares whose surfaces bore the unpredictable signs of the passage of time. In Ritsuô style lacquers, the passage of time is manipulated by being manufactured synthetically, with exaggeration deployed to make its point. Here, however, the simulated and hyperbolical signs of wear, in combination with miniaturization, self-consciously direct attention to facture in a way that enhances the *inrô* not as a functional container, but as a spectacular demonstration of artistic virtuosity, displayed according to the passing whim of the fashionable consumer. The layering of past and present becomes a process of fictionalization to be controlled and endlessly revised both by the artist and by the consumer.

Acknowledgements

My thanks to Richard Checketts, Terry Milhaupt and, especially, Andrew Watsky for their thoughtful comments and suggestions for revisions to this essay.

Notes

1 My thanks to Terry Satsuki Milhaupt for suggesting the analogy between the actual layering of the *jûnihitoe* and the technical elaboration of the *kosode*. For a detailed analysis of the most elaborate of these techniques see her 'Flowers at the crossroad: the four-hundred-year life of a Japanese textile', PhD Dissertation, Washington University, St Louis, 2002.

2 It is reproduced in D. C. Gluckman and S. Sadako Takeda, *When Art Became Fashion: Kosode in Edo-period Japan* (Los Angeles: Los Angeles County Museum of Art, 1992), p. 50.

3 On layering in textiles see Nagasaki, S., *Kasane No Irome* (The combination of colours in layers) (Tokyo: Tankôsha, 1987); L. Crihfield Dalby, *Kimono: Fashioning Culture* (New Haven and London: Yale University Press, 1993); and C. Cavanaugh, 'Text and textile: unweaving the female subject in Heian writing', *Positions*, 4:3 (Winter 1996), 595–636.

4 P. Bourdieu, *Distinction: A Social Critique of the Judgment of Taste* (R. Nice trans.) (Cambridge: Harvard University Press, 1984).

5 For a discussion of the range of articles made from lacquer and the attitudes towards this material, see A. M. Watsky, *Chikubushima: Deploying the Sacred Arts in Momoyama Japan* (Seattle: University of Washington Press, 2004), especially pp. 143–196.

6 J. Earle, *Japan. Splendors of Meiji Japan: Arts of the Meiji Period from the Khalili Collection* (London: The Khalili Family Trust, 2001), p. 210.

7 A. Heginbotham and M. Schilling, 'New evidence for the use of South-East Asian materials in seventeenth-century Japanese export lacquer', paper presented at 'Crossing borders:– the conservation, science, and material culture of East Asian lacquer', Victoria and Albert Museum, 30–31 October 2009.

8 For examples of surviving sets see *Konrei* (Trousseau Heirlooms of Daimyô Ladies), Treasures from the Tokugawa Museum No. 7 (Nagoya: Tokyo Museum, 1991).

9 For a good description of the techniques with particular emphasis on the qualitative differences between strata, see B. T. Okada, *Symbol and Substance in Japanese Lacquer: Lacquer Boxes from the Collection of Elaine Ehrenkranz* (New York and Tokyo: Weatherhill, 1995), pp. 13–21.

10 A. Yonemura, 'Splendour and supremacy: lacquer for the Tokugawa', in *Japan of The Shoguns* (Montreal: The Montreal Museum of Fine Arts, 1989), p. 78.

11 *Konrei*, p. 140.

12 The diary is translated in A. J. Pekarik, *Japanese Lacquer, 1600–1900: Selections from the Charles A. Greenfield Collection* (New York: The Metropolitan Museum of Art, 1980), pp. 121–123.

13 D. H. Shively, 'Sumptuary regulation and status in Early Tokugawa Japan', *Harvard Journal of Asiatic Studies*, 25 (1964–65), 144.

14 Pekarik, *Japanese Lacquer*, citing an edict of 1724, p. 43.

15 E. Ikegami, *Bonds of Civility: Aesthetic Networks and the Political Origins of Japanese Culture* (Cambridge: Cambridge University Press, 2005), p. 256 citing an edict of 1648 and Shively, 'Sumptuary regulation and status', 155.

16 J. C. Y. Watt and B. Brennan Ford, *East Asian Lacquer: The Florence and Herbert Irving Collection* (New York: The Metropolitan Museum of Art, 1992), p. 175.

17 J. Tanizaki, *In Praise of Shadows* (T. J. Harper and E. G. Seidensticker trans.) (New Haven: Leete's Island Books, 1977), p. 11.

18 Y. Kenkô, *Essays in Idleness* (D. Keene trans.) (New York: Columbia University Press, 1998), p. 82. It is not clear from the original, in which the term *karabitsu*, or 'Chinese chest', is used whether or not it was lacquered. *Hôjôji* (The ten foot-square hut), *Tsurezuregusa* (Essays in idleness) in *Nihon Koten bungaku taikei* (Compendium of classical Japanese literature) 30 (N. Minoru ed.) (Tokyo: Iwanami Shoten, 1957), p. 169.

19 The first reference to Negoro ware dates from 1638, according to S. Kawada, *Negoro* (Kyoto: Shikôsha, 1985), p. 282.

20 On various views of the origins of the designation, see *ibid.*, pp. 282–284.

21 S. Greenblatt, 'Resonance and wonder', in I. Karp and S. D. Levine (eds), *Exhibiting Cultures: The Poetics and Politics of Museum Display* (Washington, D.C.: Smithsonian Institution, 1991), p. 44.

22 Although tea aesthetics contributed to the canonization of Negoro, it was a relatively late development and these wares themselves do not figure very prominently in *chanoyu*. Those few examples recorded in tea diaries include trays for sweets and

containers for incense. See S. Hayashiya (ed.), *Cha no dôgu* (Tea utensils, part 2), in *Chadô shûkin* (A brocade of tea), vol. 11 (Tokyo: Shogakkan, 1983), pp. 203, 237, 240.

23 Quoted in T. Haga, 'The Wabi aesthetic', in P. Varley and I. Kumakura (eds), *Tea in Japan: Essays on the History of Chanoyu* (Honolulu: University of Hawaii Press, 1989), p. 197.

24 See *Higashiyama Gyobutsu*, Nezu Museum and Tokugawa Museum, comp. (Tokyo, 1976), pp. 221–2.

25 For a discussion and examples, see Watt and Brennan Ford, *East Asian Lacquer*, pp. 26–39.

26 G. McCracken, *Culture and Consumption: New Approaches to the Symbolic Character of Consumer Goods and Activities* (Bloomington and Indianapolis: Indiana University Press, 1990), pp. 39, 35.

27 See H. D. V. Prendergast, H. Jaeschke and N. Rumball, *A Lacquer Legacy at Kew: The Japanese Collection of John J. Quin* (London: Royal Botanic Gardens, Kew, 2001), no. 146.

28 On Ritsuô, see Pekarik, *Japanese Lacquer*, pp. 84–91.

29 On *inrô* see J. Hutt, *Japanese Inrô* (Tokyo: Weatherhill, 1997).

30 Translated by Pekarik in *Japanese Lacquer*, p. 124.

31 *Ibid.*, pp. 124–126.

32 The Victoria and Albert Museum also has in its collections a round writing box by Ritsuô from 1746 made of lacquer ceramics, ivory and shell that simulates the appearance of a worn temple gong. See http://collections.vam.ac.uk/item/O102165/writing-box/.

33 For other interpretations of this important concept in Japanese culture see T. Clark, 'Mitate-e: some thoughts and a summary of recent writings', *Impressions*, 9 (1997), 6–27 and M. Yamaguchi, 'The poetics of exhibition in Japanese culture', in Karp and Levine, *Exhibiting Cultures*, pp. 57–67.

4 ✧ Patina and the bourgeoisie: the appearance of the past in nineteenth-century Paris

Manuel Charpy

THE history of nineteenth-century bourgeois material culture, especially in Paris, focuses mainly on the *grands magasins* (department stores) and the consumption of novel industrially made products. However, bourgeois consumption was more complex and paradoxical than that; it included many old, as well as new things. In 1852, the same year that the Bon Marché – the first grand magasin in Paris – opened, the great auction room, Drouot, a temple of old curiosities and antique goods, was also inaugurated. A new passion for antique goods developed in tandem with, and in opposition to, that for commercial and industrial novelties. The bourgeoisie sought unique and authentic goods, the antithesis of the serial industrial products offered by the grands magasins. In these antiques, another kind of novelty emerged: patina. This is a key instance of the surface as a *sémiophore* – Krzysztof Pomian's term for a sign that takes part in the formation of social identity.[1]

In the world of appearances that was the bourgeois apartment, patina was singular. In general, surfaces in these interiors were functional or orthopaedic, forcing occupants to adopt a way of life, to adopt self-control.[2] Varnished furniture, velvet that marked the body's tracks, perfectly white linen, polished silver: these surfaces forced people to clean houses continuously, or keep servants to do so for them, amounting to a denial of the wear and tear involved in everyday life – much as polished shoes did in public spaces.[3] Surfaces proclaimed the social rank of the bourgeois. In this context, patina seems to have been an anomaly, because it reflected the 'grime of the times'. How should we understand the place and the sense of this particular surface in private spaces, and especially in Parisian bourgeois apartments?

The term 'patina' came from the art world, and especially from numismatics; at the end of the eighteenth century it had the narrow meaning of the verdigris colour that copper takes on over time.[4] In 1800, Millin's monumental dictionary noted that the French language lacked the word for any aged appearance and so had to use the Italian term, *patina*.[5] In this context the term already signified 'proof of antiquity', but it only gradually came to designate the surface that paintings acquire over time. This was an accepted meaning by 1840, at which point another dictionary defined patina as the 'matter that accrues on paintings'.[6] From the 1850s onwards, while a marketplace of antique goods grew up, the term's meaning gradually extended to designate 'the nuance that time gives to *objets d'art*'.[7] Around 1850, word and thing converged in the marketplace.

Thanks to shopkeepers' archives, the bankruptcy files of antique dealers and interior decorators, auctioneers' archives, and private accounts and diaries, we can understand how the Parisian bourgeoisie invented or reinvented the patina in their apartments and how this demand created a new way to use this appearance of the past.

The birth of the antique

The passion for patina, and antiques more broadly, arose first among marginal collectors of the 1830s – Balzac's Cousin Pons for example. By the 1850s, however, the bourgeoisie had discovered and adopted these goods.[8] A marketplace developed to serve the new demand for old goods. Another etymological shift tells the story: at the beginning of the nineteenth century, *antiquités* designated Greco-Roman artefacts; after 1850, it meant all kinds of objects from the past, without distinction. Also by mid-century there was a distinction between the dealers of antiques and those of second-hand goods (*friperie et objets de brocante*); patina was one of the most important ways to distinguish these two categories.

A new site emerged where the Parisian bourgeoisie could find the patina they desired: Drouot, the auction house. Drouot hotel was built in 1852, close to the fashionable boulevards in the north of the city.[9] Already at the end of the eighteenth century, the auctions had attracted *amateurs* interested in paintings. But, by the middle of the nineteenth century, the wider bourgeoisie was looking for legislators of taste. Auctioneers – members of the legal profession in France – tried to meet this demand. They gathered all Parisian auctions in a new building and created a new scenography. Intent on conquering the public, the building – which combined the qualities of an administrative palace, a monument and a temple – sharply demarcated their goods, and hence their own public: the first floor was for second-hand goods and second-hand dealers; the second floor, with more luxurious rooms, was for antiques sold by more prominent dealers. Auctioneers' archives attest that aristocrats and bourgeois alike came to find all kinds of goods: trinkets and curios, paintings, furniture and architectural fragments, all of which were patinated. From its opening onwards, Drouot enjoyed

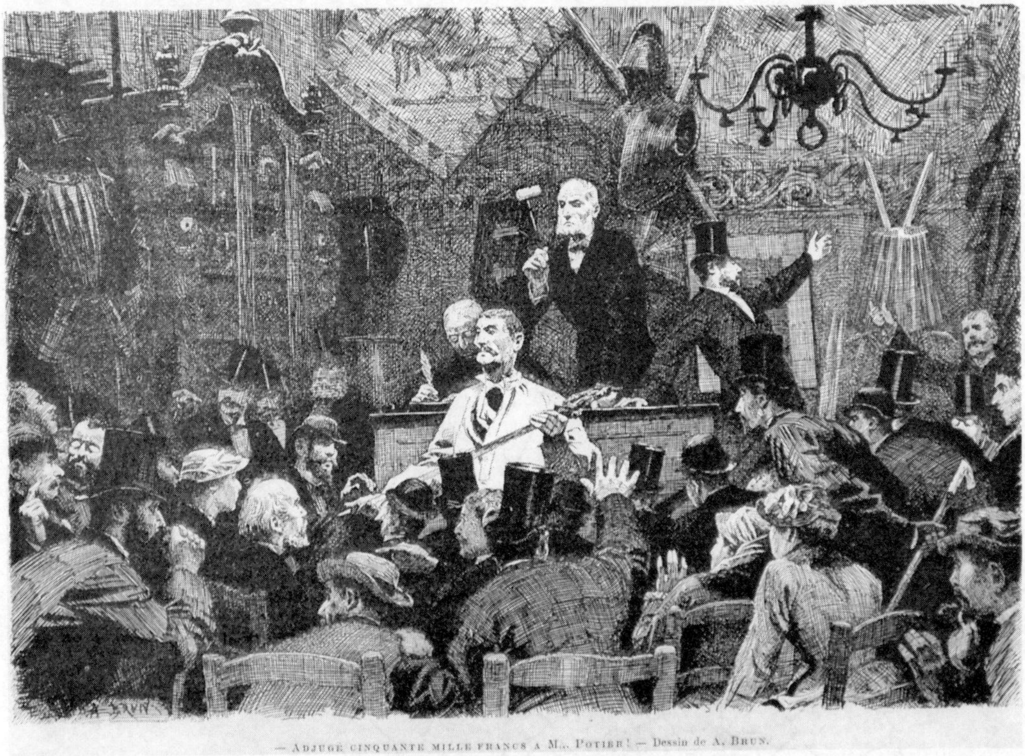

9 From a drawing by A. Brun, auction of antique goods at Hotel Drouot, with a bourgeois audience, c. 1880

great success. The *Journal Illustré* wrote: 'The auction room was, for a few years, a crowded building. Everybody gathered there. Buyers and sellers filled the place, and it was there that the great shambles that one calls Paris was cleared off.'[10] There, clients found authentic, unique goods guaranteed by experts, with singular stories of provenance told in catalogues, or by auctioneers. At Drouot, the bourgeois learned about the price and cultural value of antique goods. The journalist Octave Uzanne noted: 'Drouot is our Museum of arts and crafts.'[11] Drouot was one of the places where the cult of patina was established as a seal of authenticity (Figure 9).[12]

Around 1850, there were 25 antique dealer's shops in Paris; by 1860 the number had grown to 200, and the figure reached 300 in 1890 (Figure 10). These were located mostly in two areas of the city: the newly urbanized, fashionable districts of the northern boulevards and Saint-Georges quarter (where many artists lived); and le Vieux Paris (Old Paris), areas that had escaped the reconstruction works of Baron Haussmann, around Boulevard Beaumarchais, Notre-Dame, La Sorbonne and, above all, along the Seine

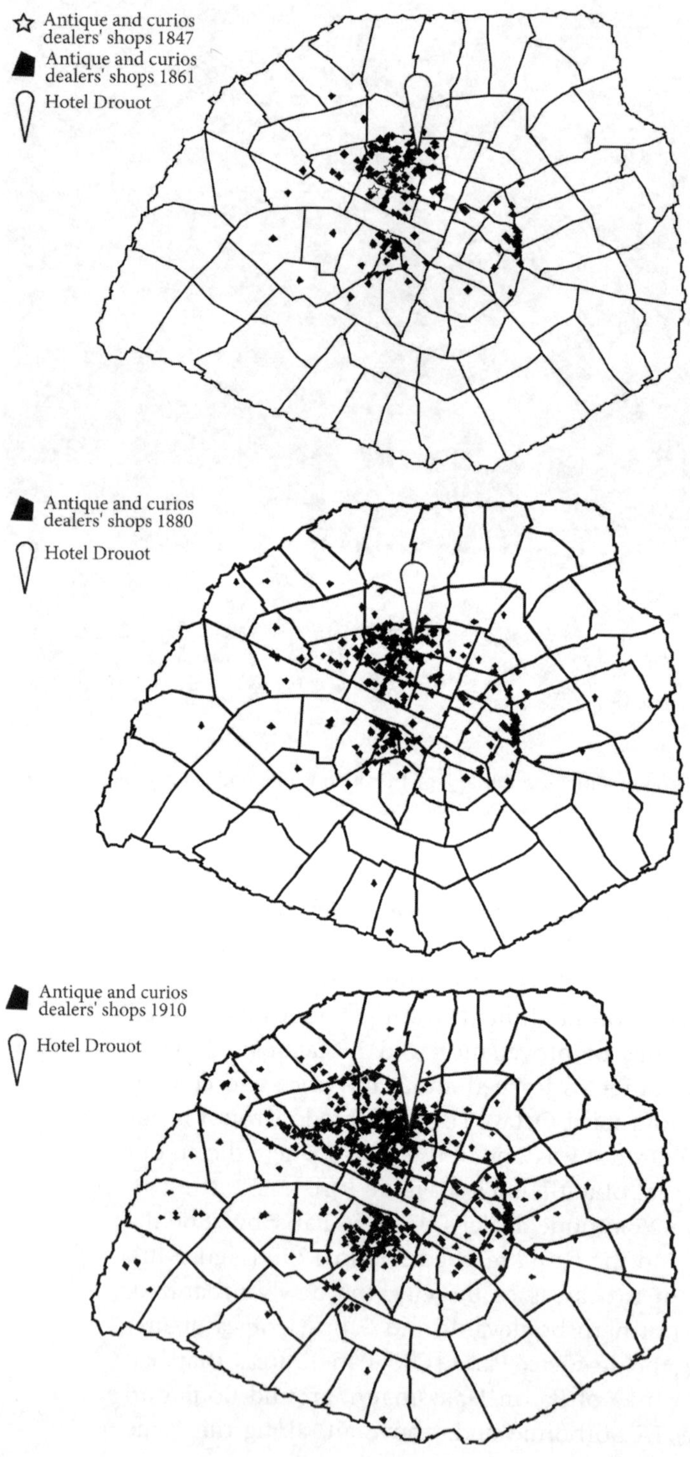

10 Hotel Drouot and antique dealers' shops in Paris.

river banks. With the urban mutations of the 1860s, a new nostalgic vision of Paris was developing. These spaces of *bibelotage* (antique hunting) were areas of promenade and tourism, in other words areas of travel into the past. Antique goods dealers fed this imagination: they made new-old façades with half-timbering, hung old signs and invented suggestive names such as 'Au Vieux Paris' and 'Souvenirs d'antan' (Figure 11).

After 1860, publishers began to exploit the new interest, bringing out handbooks of *savoir-faire* that advised neophyte collectors where and how to choose antiques.[13] These texts stressed the importance of patina in valuing goods, in particular rustic and exotic items, and objects of 'human antiquity'. Buying antiques required an education of the hands as well as the eyes (Figure 12). Touching an old object was the way to read, or decode, the patina. The surface was the terrain in which a new social ability was proven, leading to the foundation of a new visual and tactile culture.

The range of antiques found in a bourgeois apartment of the time was extremely broad: not only European goods from the fifteenth to the end of the eighteenth centuries, but also exotic things. During the nineteenth century, the craze for goods from the Middle East, China, Japan, pre-Columbian America and the Pacific Islands reached the bourgeoisie, who were fascinated by traces of civilizations that appeared as 'dying civilizations' or civilizations from the past. By 1870, there were more than 200 dealers of curios (*curiosités*), specialized in all kinds of exotic goods.[14] The

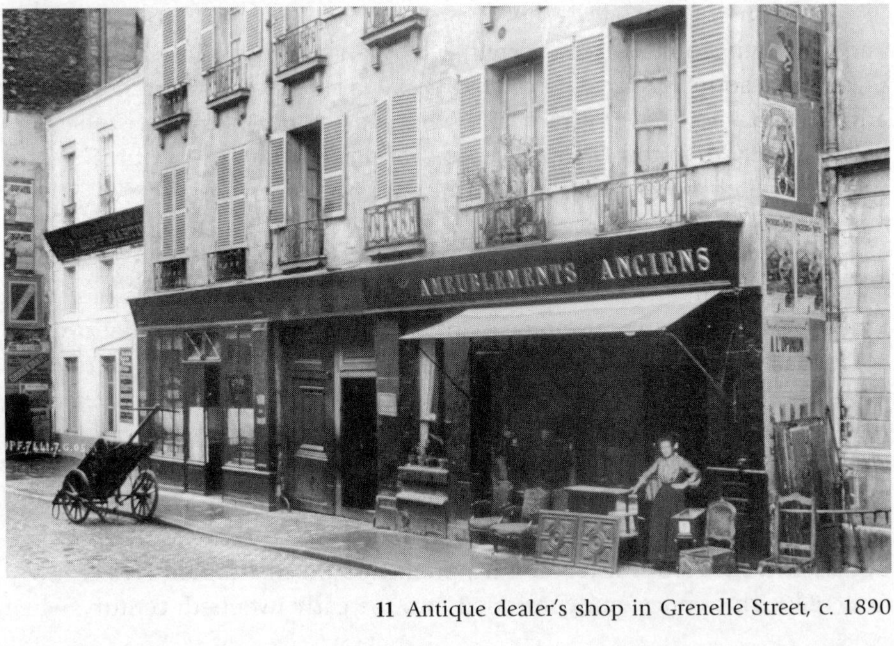

11 Antique dealer's shop in Grenelle Street, c. 1890

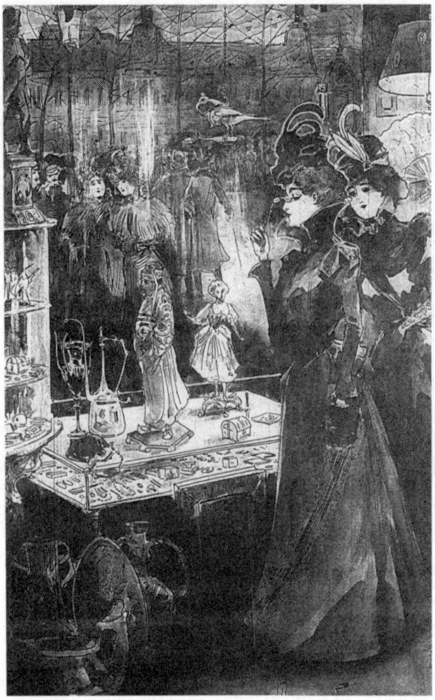

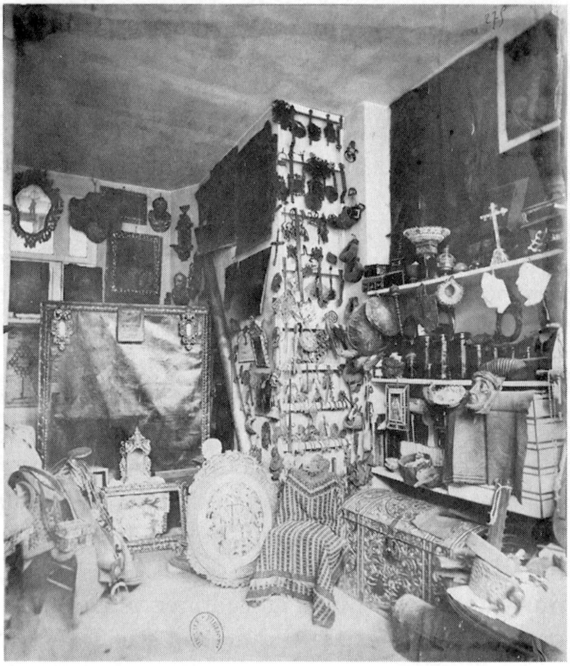

12 Gaston Noury, bourgeois women in an antique dealer's shop on banks of the Seine, watercolour, 1897

13 The shop of antique dealer Eugène Boban, boulevard Saint-Germain, Paris, c. 1870–80

Parisian bourgeois also discovered with delight the rustic objects of the French peasantry, covered with traditional ornament. And with scientific advances in knowledge of prehistory, buyers started to buy old stones and fossils from the 'mists of time'. These various classes of antique – exotic, rustic and prehistoric – all had a common sign: patina (Figure 13).

The new science of patinas: from description to production

One of the central functions of early art history was its formulation of a 'science of patinas', an ecumenical process in which experts and auctioneers authenticated paintings alongside *objets d'art* and furniture. Similarly, archaeologists attempted to date objects from the ground and in particular bronzes from Antiquity on the basis of patina – a process that was soon aided by chemists, who tried to understand the phenomenon from a technical point of view and classify all types of the natural traces of time. More than fifty books, and numerous articles, were published on this topic between 1860 and 1880. This endeavour continued into the early twentieth century, when one chemist wrote: 'We aim to seek out by microscope the tracks and the nature of the intimate phenomenon of patina.'[15]

Right from the beginning, this research inadvertently fuelled a phenomenon that was in some sense its antithesis: the industrial production of patinas. Paintings were the first to receive the simulated effect of ageing, with the invention of a new kind of varnish that created artificial *craquelure*. The second half of the century was the great moment of the production of copies, across a continuum from the frankly deceptive to the merely evocative. Painters working in small shops, daubers (*rapin*), copied old paintings from the seventeenth and eighteenth centuries, as well as (after 1880) family daguerreotypes. The bourgeoisie could now obtain images of their own forefathers in oil painting, a technique more prestigious than photography. Of course, these paintings were patinated.[16] With the increase of amateurism, dealers found a new market for their varnishes and paints that imitated the appearance of the old.[17] Amateurs, in particular women, learnt to produce patina on domestic goods, especially on sentimental souvenirs: frames for portraits, family sculpture busts, jewel boxes, etc. While hygiene and cleanliness still dominated in the bourgeois home, manufacturers began offering a wax in the 1880s that artificially aged the surfaces of furniture.[18] Similar effects were being achieved industrially. Thanks to such innovations as electroplating, and synthetic varnishes and polishes, all kinds of products could be made to take on the appearance of the past.[19] Goods made of wood, metal, papier mâché, gutta-percha (a plant-based material similar to rubber) or *bois durci* (sawdust mixed with oxblood) could all masquerade as products of the Ancien Régime, the pre-industrial world of crafts. Nearly a hundred patents for such processes were registered between 1860 and 1900.[20] Even the French language recorded the change: there was now a verb, *patiner*.[21]

False patina: making the traces of past

It is a fine line between imitation and falsification, and as can well be imagined, the widespread practice of patination included a great deal of outright faking. Forgers quickly grasped the role of patina in securing authenticity, and began developing a repertoire of false surfaces from the 1850s onwards. A lawsuit that took place in 1858 helps us to understand the new place of patina in bourgeois material culture. Boissel de Monville, an aristocratic expert who had worked for the Rothschild brothers, bought many fake antique pieces (supposedly sixteenth century in date) from Pierrat, 'a restorer of *objets d'art*'. The prosecutor noted that Pierrat told many false stories about the objects and sold them in strange and picturesque places in order to deceive the customers. But, above all, he deceived Boissel with false patina. The prosecutor accused him in no uncertain terms: 'In order to sell these novelties as antiques, you coat their surfaces with dirt, dust and varnish. You play upon amateurs' credulity.' Boissel's lawyer underlined the importance of patina, saying to the forger: 'You imitated antiques … It's not forbidden. But you covered them with a varnish in order to change the colour, the appearance of them … here is the crime!' Asked whether the surfaces on these objects might simply be dusty, rather than intentionally deceptive, Boissel answered: 'No. Ordinarily, when objects are dusty from a workshop, water and soap are sufficient. But I rubbed this grime for four hours! Soap, caustic potash and ether did nothing. It is said that forgers use a varnish and then heat the fakes in a kiln.'[22]

Boissel was right – this was one means by which fakes were manufactured, but hardly the only way. A market for fakes had already existed prior to the popularization of antiques during the 1850s and 1860s, but it now became pervasive. Nestor Roqueplan, a journalist and collector, wrote in 1869:

> It is impossible to eat, drink, sit, to live at all without inhabiting the history of France in wood, glass or porcelain […] Unless there is a subterranean or airborne France where all the old pieces of furniture from the monarchy have been kept, how is it possible that in just ten years, all modern France has been furnished, as if by magic, with our forefathers' antiques? […] There is no trade where cheating is practised with such effrontery.[23]

Another journalist, Henri Rochefort, noted that: 'The besmircher (*salisseur*) of paintings knows the human heart […] Bone charcoal, liquorice juice and ashes in water are blended over the painting; having bought a painting in Drouot, cleaned to look brand new, it is made old once again.'[24]

In the case of furniture, the production of fakes was especially complex. Forgers used new chemicals to work more quickly, creating verdigris in a few hours using hydrofluoric acid, eating away at wood surfaces using nitric acid, and fading the colour of timber with potassium permanganate.[25] The prolific writer and collector Paul Eudel conducted an inquiry about the industry in 1880. He wrote that:

> For twenty years, craftsmen in Auvergne have been covering new pieces of furniture with … many coats of paint. Then, they expose them to the sun. After one or two years, they scratch the paint off with caustic potash. It makes wonderful panellings and pieces of furniture that testify their age with the tracks of paint in the wood's cracks. Then, they bring their products to Paris … They are moving back the hand of time.[26]

Most of the fake industries were settled in Paris, but in some cases forgers produced fakes and then left them in villages and inns for bourgeois buyers to find 'by chance' while on holiday.[27] This was especially common along the road between Paris and Normandy (Figure 14).

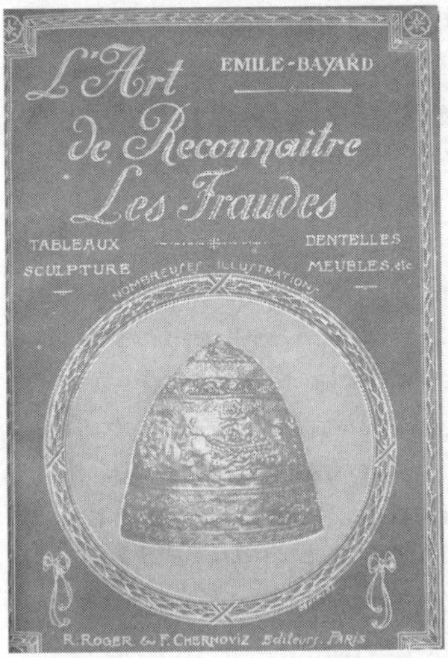

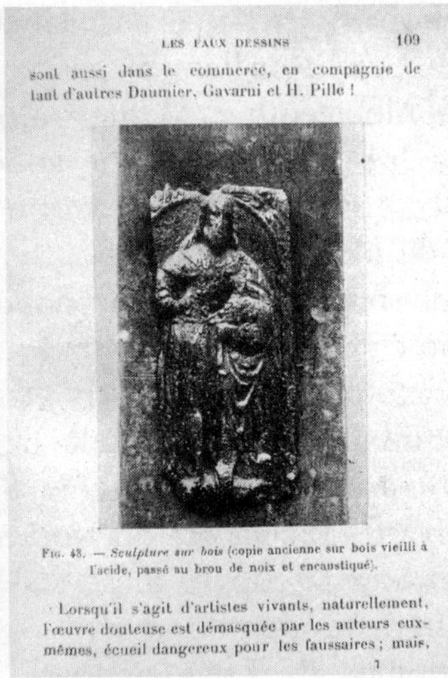

LES FAUX DESSINS 109

sont aussi dans le commerce, en compagnie de tant d'autres Daumier, Gavarni et H. Pille !

Fig. 48. — *Sculpture sur bois* (copie ancienne sur bois vieilli à l'acide, passé au brou de noix et encaustiqué).

Lorsqu'il s'agit d'artistes vivants, naturellement, l'œuvre douteuse est démasquée par les auteurs eux-mêmes, écueil dangereux pour les faussaires ; mais,

7

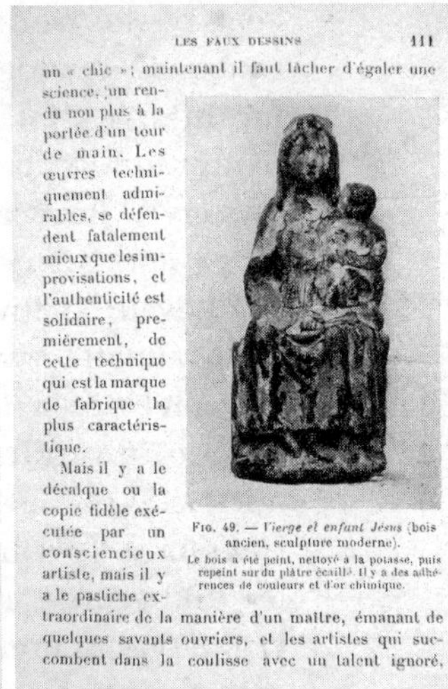

LES FAUX DESSINS 111

un « chic » ; maintenant il faut tâcher d'égaler une science, un rendu non plus à la portée d'un tour de main. Les œuvres techniquement admirables, se défendent fatalement mieux que les improvisations, et l'authenticité est solidaire, premièrement, de cette technique qui est la marque de fabrique la plus caractéristique.

Mais il y a le décalque ou la copie fidèle exécutée par un consciencieux artiste, mais il y a le pastiche extraordinaire de la manière d'un maître, émanant de quelques savants ouvriers, et les artistes qui succombent dans la coulisse avec un talent ignoré,

Fig. 49. — *Vierge et enfant Jésus* (bois ancien, sculpture moderne).
Le bois a été peint, nettoyé à la potasse, puis repeint sur du plâtre écaillé. Il y a des adhérences de couleurs et d'or chimique.

14 Émile Bayard, *L'Art de reconnaître les fraudes*, Paris: R. Roger et F. Chernoviz, 1914

The professional make-up artist

André Mailfert, based in Paris and Orléans, was one of the most important forgers in France at the beginning of the twentieth century. Though perhaps best known for having completely invented an eighteenth-century cabinetmaker (named Jean-François Hardy) and an entire stylistic school (l'École de la Loire), his main activity was the production of fake antique furniture. He invented numerous processes for yellowing ivory, ageing wood, etc., which were recorded in his memoirs titled *Au pays des antiquaires: confidences d'un 'maquilleur' professionnel* (*In the Land of the Antiquaries: Secrets of a Professional 'Make-up Artist'*). Describing a craquelure effect achieved with an electric heater and a special varnish, which 'added a sign of authenticity and real delicacy', Mailfert emphasized that this work must be done 'with perspicacity, because patina is above all an art of observation and psychology'. Patina has to reproduce not only the effects of time but also the marks of previous owners. Thus, for example, the surfaces around handles, feet and armrests had to be worn away, then treated with wax and dust. However despite the care he claimed to lavish on his productions, Mailfert operated at industrial scale. In his factory – a factory of the past – he employed two full-time dust collectors. (According to Mailfert, Louis XV furniture was fashionable at the time and its carved surfaces required a great deal of dust.) The earth used by many forgers, he wrote, didn't look convincing. His assistants went to the extent of gathering their dust from churches, where it had accumulated since the twelfth century. It was then kept in the workshop in ten large jars, labelled by era. This methodical craftsman catalogued eight degrees of patina, from the twelfth century to the middle of the nineteenth.[28] Mailfert also sometimes smoked pieces of furniture over a fire, to 'envelop the goods with the essential veil of mystery'.[29] Ironically, just after the First World War, he sold boxes of ready-to-use 'antique dealer patina'.[30]

Private archives show that by the end of the century, artisans were also patinating whole interiors. For example, between 1880 and 1910, the apartment of a Doctor Richelot in Paris was entirely resurfaced to look old: not just furniture and panelling, but the fireplace, stairs and *trumeau* mirrors.[31] In new apartments built during the Second Empire all along Haussmann's boulevards, the decorators worked to conceal the recent rise of the bourgeoisie. The new façades of these buildings looked like old palaces. White plaster and new wooden floors were covered with old tapestries and carpets, and stained glass filled the windows.

By 1880, this general aesthetic of 'dilapidation' (*vétusté*), or glorious decay, was well established among the bourgeoisie. This taste was indissociable from patina. Dealers bought old wood and old tapestries – Mailfert sent his employees to the countryside and to urban second-hand markets in order to find old beds and wood panelling – but with the increase of demand, the supply of antiques was not sufficient.[32] Once again forgers invented new process to reproduce the effects of time. For panelling and wooden floors, they substituted modern dance floors. Around 1880, they developed an electric machine that bored imitation wormholes.[33] Holes, cracks and marks were picturesque signs of history; they told the story of the goods, a great adventure through time. Paul Eudel visited one such 'factory of old furniture' (in his paradoxical expression) and described the scene: 'a specialised crew of workers was armed with big sticks … Men hit hard. They punished pieces of furniture for their newness. In this way, they made holes, scratches and impacts that an old item would accumulate over a long life.'[34] Twenty years later, Mailfert and his employers were using a more diverse toolkit: 'metal combs, wooden hammers, whips with hooks, etc. I made a whip with bolts, old keys, screws, hooks and many various trinkets. We whipped pieces of furniture with it; every knock increased the value.'[35] The Parisian bourgeois left Paris and scoured the provinces, and also went abroad, in order to escape the fakes. But forgers were expanding just as fast, into new product lines such as false prehistoric flints.[36]

This new passion for patina was quickly known worldwide. Indeed, it was a time of great increase in bourgeois tourism, within Europe and in the East. As the archives of Eugène Piot – a collector and traveller – show, dealers and craftsmen sold antique goods as souvenirs. In the bourgeois mind, a true souvenir required patina because travelling in space meant travelling into the past as well. A souvenir had to reflect local colour. Just one example among hundreds: when Paul Eudel arrived in the grand bazaar of Constantinople (Istanbul), he noted that all the craftsmen 'aged' goods with acids, dust and wax for European tourists. Even when tourists saw them at work, he noted, five minutes later they bought these 'old antiques produced this very day'.[37] The same was happening in Venice, Rome, Cairo, Algiers, and as far away as Beijing and Canton.[38] For the bourgeois customer, all objects produced by 'exotic' craftsmen were authentic goods, because the artisans themselves came from the past; they belonged to tradition. So, with the increase of tourism, the new culture of patina spread around the world, and changed the traditional production of goods. The western imagination colonized the material culture of the world.

Conclusion: narratives of authenticity

Why was this surface – whether false or true – so successful? The first reason is that patina lent distinction. This was true across the wide spectrum captured by the term 'bourgeoisie', from the inherited heirlooms of the nobility to inexpensive fakes bought by the lower middle class. With patina, the bourgeoisie took possession of time itself. It was a way to appropriate the past, through sensibility and imagination. The bourgeoisie's rise came through industry and trade, and collecting patinated goods was also a way to claim cultural and historical legitimacy. Antique (and seemingly antique) goods connected the bourgeois interior with the world before the French Revolution and the industrial revolution. In this way, the bourgeois took on the role of an heir in an old family. Patina both distinguished the bourgeoisie from the lower class and, at the same time, hid its recent rise.

A further reason for this surface's success was its authentic character. Paradoxically, the bourgeoisie thought that its own world – urban, industrial and commercial – was artificial, a fake world. Patinated goods, on the contrary, were like natural products. They seemed to come from traditional society, before economic and political revolutions. They were considered to be unique goods *par excellence*, and therefore an escape route from mass-produced copies.

The third reason for the success of patina concerned bourgeois relations to time and the past. As Walter Benjamin wrote: 'The private individual, who in the office has to deal with reality, needs the domestic interior to sustain him in his illusions … In the interior, he brings together the far away and long ago.'[39] Indeed, domestic apartments became anachronistic spaces, and retreats from the modernity of urban city life. In the nineteenth century people learned a new temporal arithmetic; public space, in particular, was now subdivided into minutes and seconds.[40] The more nebulous time of patina was reassuring, allowing unlimited reverie within the bourgeois private space. As Aloïs Riegl wrote in his 1903 book about the modern cult of ancient monuments, the nineteenth century had witnessed a shift from 'historical value' (the monument as a document) to 'age value'.[41] The same change occurred in private space. Consumers sought patina as a pure sign of the past, without necessarily needing a link with real history or events. The past captured in patina was smooth, without specific accidents, revolutions or disruption.

The final reason for the popularity of patina was that it fed the bourgeois imagination. Right from the beginning of the cult of antiques, around 1830, the taste was adopted by artists, in particular painters, who bought themselves props to aid in creating archaeologically correct motifs. Though patina and dilapidation were generic, they could be used to tell a singular story. The bourgeoisie in their apartments were like the viewers of academic paintings or readers of historical novels: a few elements or appearances of reality were sufficient to believe the story. Émile Bayard, a Beaux-Arts inspector wrote around 1900: 'that is the merit of the antique: the softness of its forms has been caressed by the dust of time, the warmth of its colours mysteriously enveloped in the fog of the past'.[42] Patina told stories because it brought out local colour. It recorded, in the imagination, the accumulated atmosphere of different eras that the object had inhabited. As Aloïs Riegl noted, it was the same with ancient monuments: 'Age value [in a monument] is revealed in imperfection, a lack of completeness, a tendency to dissolve shape and colour, characteristics that are in complete contrast with those of modern, i.e., newly created, works.'[43] It was this 'strange colour' of the old stone that moved the tourist.

In an essay published in 1907, sociologist Georg Simmel analysed the ambiguous pleasure arising from the contemplation of ruins. He wrote that these sites fulfilled two contradictory desires: the desire to experience history – 'the shadow of nostalgia' – and the desire to get out of history altogether.

> The ruin creates the present form of a past life, not according to the contents or remnants of that life, but according to its past as such. This is also the charm of antiquities … No matter if we are deceived in an individual case: with this piece which we are holding in our hand, we command in spirit the entire span of time since its inception; the past with its destinies and transformations has been gathered into this instant of an aesthetically perceptible present.[44]

Patina could be considered as an adaptation to private space of this aesthetic. It was a picturesque surface; certainly the most romantic surface in the bourgeois interior, and perhaps the most bourgeois as well.

Notes

1 K. Pomian, 'Histoire culturelle, histoire des sémiophores', in *Sur l'histoire* (Paris: Gallimard, 1999), p. 255; D. Miller (ed.), *Material Cultures: Why Some Things Matter?* (Chicago: University of Chicago Press, 1997).

2 See V. Kelley, *Soap and Water: Cleanliness, Dirt and the Working Classes in Victorian and Edwardian Britain* (London: I. B. Tauris, 2010); V. Kelley, 'Housekeeping: shine, polish, gloss and glaze as surface strategies in the domestic interior', in D. McMahon and J. Myers (eds), *Material Possessions: The Objects and Textures of Everyday Life in Imperial Britain* (Athens, Ohio: University of Ohio Press, forthcoming). See also, M. Charpy, *Le Théâtre des objets: espaces privés, culture matérielle et identité bourgeoise, Paris, 1830–1914*, PhD Dissertation, Université François-Rabelais de Tours, 2010, pp. 309–341.

3 See T. Veblen, *The Theory of the Leisure Class: An Economic Study in the Evolution of Institutions* (New York and London: Macmillan, 1899).

4 See A. Rey, 'Patine', in *Dictionnaire historique de la langue française* (Paris: Le Robert, 1998). One of the first chemical studies about patina was published during the 1780s: L. Bernard Guyton de Morveau, H. Maret and J. P. François Guillot Duhamel, 'Patine', in *Encyclopédie méthodique: chymie, pharmacie et métallurgie*, vol. 5 (Paris: Panckoucke, 1786).

5 'Patina', A.-B. Millin, *Dictionnaire des Beaux-Arts* (Paris: Imprimerie Crapelet, 1806).

6 Rey, 'Patine'.

7 W. Duckett, *Dictionnaire de la conversation* (Paris: Firmin Didot frères, 1855).

8 See Charpy, *Le Théâtre des objets*, chapter 5, 'Authentiques antiquités: usages domestiques du passé et invention d'une marchandise'. On the same phenomenon in the United Kingdom, see M. Westgarth, *The Emergence of the Antique and Curiosity Dealer 1815–1850: The Commodification of Historical Objects* (Farnham: Ashgate, 2011).

9 M. Charpy, 'The auction house and its surroundings: the trade of antiques and second-hand items in Paris during the nineteenth century', in B. Blonde, J. Stobart and N. Coquery (eds), *Between Novelties and Antiques: Mixed Consumer Patterns in Western European History* (Turnhout: Brepols, 2008), pp. 217–233.

10 'L'Hôtel Drouot', *Journal Illustré*, 29 April – 6 May 1866. All translations are mine unless otherwise indicated.

11 O. Uzanne, 'L'Hôtel Drouot et la curiosité', in *Les Zigzags d'un curieux* (Paris: Quantin, 1888).

12 See Charpy, 'The auction house' and Charpy, *Le Théâtre des objets*, chapter 9, 'L'invention de Drouot: nouvelle institution du goût et de l'authenticité'.

13 See P. Eudel, *Le Truquage, altérations, fraudes, contrefaçons dévoilées* (Paris: Editions Rouveyre, 1884); E. Bayard, *L'Art de reconnaître les fraudes* (Paris: R. Roger et F. Chernoviz, 1914); A. Blanchet and F. de Villenoisy, *Guide pratique de l'antiquaire: nettoyage, restauration, conservation et reproduction des objets archéologiques* (Paris: Leroux, 1899).

14 See 'Marchands de curiosités' and 'Antiquaires' in *Annuaires Bottin-Didot du commerce*, 1868–90.

15 *Le Bulletin de la Société normande d'études préhistoriques* (1906), 95. Many studies on patina were published at the end of the century in artistic and archaeological reviews and fascicule form, see for example, F. de Villenoisy, *La Patine du bronze*

antique (Paris: E. Lerous, 1896) (extract from the *Revue archéologique*, 1896) or Henri Lechat, *La Patine des bronzes grecs* (Paris: E. Leroux, 1896.) In the same time, several historical, archaeological and prehistorical societies tried to describe the process of formation and the scientific reviews wanted to understand too: see for example, *La Revue scientifique* on the formation of 'Genuine patina of Japanese goods' (1895), 220 and *Annales de chimie et de physique* on 'Green patina', 11 (1897), 443.

16 See for example, the huge file about these practices: Archives de Paris, D2U6/91, 'Comptoir du progrès universel (portraits en peinture et dessin à partir de photographies)', 1890.

17 See for example, F. Michot, *Nouveau manuel completde modelage, de moulage et de patine* (Paris: Encyclopédie Roret, 1913 [first publication 1884]) and the numerous catalogues of the Parisian house Bourgeois, the main fine art supplier at the end of the nineteenth century and at the beginning of the twentieth century.

18 See André Mailfert, *Au pays des antiquaires: confidences d'un 'maquilleur' professionnel* (Paris: Flammarion, 1935).

19 See the furniture catalogues of *grands magasins* (department stores), for example the huge catalogue of Au Bûcheron, located in the rue de Rivoli between 1870 and 1910.

20 See the archives of INPI (Institut National de la Protection Industrielle), Paris, section 'chimie' (chemistry), and *Description des machines et procédés pour lesquels des brevets d'invention ont été pris sous le régime de la loi du 5 Juillet 1844* (Paris: Imprimerie Impériale and then Nationale, 1860–1900) (chemical section).

21 Indeed, throughout the century, one can find warnings about fake patinas on bronzes. See for example 'Bronze', in *Dictionnaire universel théorique et pratique du commerce et de la navigation* (2 vols, Paris: Guillaumin, 1859–61) and 'Bronze' in A. Raffalovitch and Y. Guyot, *Dictionnaire du commerce, de l'industrie et de la banque* (Paris: Guillaumin, 1901). The latter indicated that Rome was the main place where people 'oxidize hard modern bronzes' in order to sell them as antique goods.

22 *Le Droit: journal des tribunaux, de la jurisprudence, des débats judiciaires et de la législation*, 15 October 1858.

23 N. Roqueplan, *La Vie Parisienne* (1857) (Paris: Michel Lévy Frères, 1869), chapter 20. 'Le Bric-à-brac', pp. 161–165.

24 H. Rochefort, 'Les Salisseurs de tableaux', in *Les Petits Mystères de l'hôtel des ventes* (Paris: Jules Ruff et Cie, 1863), pp. 43–45.

25 *Ibid.*, pp. 43–45.

26 Eudel, *Le Truquage*, p. 305.

27 See Roqueplan, *La Vie Parisienne*; Mailfert, *Au pays des antiquaires*.

28 Mailfert, *Au pays des antiquaires*, p. 91 and archives of the Maison Mailfert-Amos. André Mailfert reported that in 1925 he received a proposition from an American to patinate a castle in Illinois (p. 219).

29 Mailfert, *Au pays des antiquaires*, p. 41.

30 Private Archives Amos-Mailfert. A patent seems to have been recorded during the 1920s.

31 Archives nationales, Paris, AB/XIX/4226/Dossier 1, 'Factures de tapissiers-décorateurs et d'antiquaires' (interior decorators' and antique goods dealers' invoices), 1883–1913.

32 Mailfert, *Au pays des antiquaires*, p. 91. We can find many traces of this business in the

auction archives in Archives départementales of Tours, Orléans and Rouen. See for example the Archives départementales de la Seine Maritime, Commissaires-priseurs. In Bolbec, see particularly the second-hand goods dealers Prunet et fils (6E10/36 / Bolbec, archives André Bouic commissaire-priseur, 1899) and Levesque and Ridel (6E9/1 / Bolbec, répertoire des ventes et des clients, 1893–1906, Georges Pattey commissaire-priseur); in Archives départementales de Rouen, see the second-hand goods dealers Levesque, Duval and Langlois (6E6/813, Me Hurel, 1912).

33 Eudel, *Le Truquage*, p. 305; Mailfert, *Au pays des antiquaires*, pp. 29–30.

34 Eudel, *Le Truquage*, p. 305.

35 Mailfert, *Au pays des antiquaires*, p. 40.

36 See for example, Bayard, *L'Art de reconnaître les fraudes*, p. 250. See also A. Vayson de Pradenne, *Les Fraudes en archéologie préhistorique avec quelques exemples de comparaison en archéologie générale et sciences naturelles* (Paris: Nourry, 1932).

37 Eudel, *Le Truquage*, pp. 360–361.

38 See for example, articles about china ('Chine', 'Canton', 'Pékin') and pottery in Raffalovitch and Guyot, *Dictionnaire du commerce.*

39 W. Benjamin, *The Arcades Project* (Cambridge: Harvard University Press, 1999), pp. 8–9.

40 See A. Corbin, *Le Temps, le désir et l'horreur: essais sur le XIXe siècle* (Paris: Champs Flammarion, 1991) and in particular 'L'Arithmétique des jours au XIXe siècle', pp. 9–22; K. Pomian, *L'Ordre du temps* (Paris: Gallimard, 1984), pp. 271–273.

41 A. Riegl, *Le Culte moderne des monuments: son essence et sa genèse* (1903) (Paris: Le Seuil, 1984).

42 Bayard, *L'Art de reconnaître les fraudes*, p. 42.

43 Riegl, *Le Culte moderne*, p. 75.

44 Georg Simmel, 'Die Ruine: ein ästhetischer Versuch', *Der Tag*, 96 (1907); translated as 'The ruin', *Hudson Review*, 11:3 (Autumn 1958), 385.

5 ✧ A falsification of temporality: carpet distressing in Morocco

Myriem Naji

THE Sirwa is a marginal and arid mountainous area to the southeast of Marrakesh, where for centuries agro-pastoralist Berber men and women have respectively herded sheep and goats, and woven their clothing and furnishings. Up until colonial times, the weavings made by the Ayt Wawzgit ethnic group were sought after both inside and outside the region. Then, they entered the international market through Berber traders who brought them to Marrakesh. In 1938 the French colonial administration opened a weaving workshop in Taznakht, a town that had long been a major marketplace, attracting traders from both the north and south of Morocco. Today the town is the national market for carpets in the south of the country. Taznakht attracts producers from the mountains and the surrounding valleys, who come to sell their weavings every Friday. Elaborate weaving techniques have been known in the region for a long time, including the knotting technique, which was not widely practised in the mountainous region until the 1960s, but became the favourite local technique for floor coverings. It was only in the 1980s, after the decline of the sheep flocks due to a dire drought, that local weavers started using industrially spun wool imported from cities such as Marrakesh, Rabat or Casablanca in response to increasing international market demand.

It seems that the practice of giving textiles an antique look through lengthy sun exposure started around that time, or possibly before. According to one practitioner in Taznakht, a certain dealer in Marrakesh noticed that his customers preferred the tattered old carpets he had displayed outside his shop to attract the passers-by to the new ones he had inside the shop. The production of outright fakes seems to have started in the 1990s, when Moroccan textiles were suddenly given collectable status after several western dealers and collectors organized various events to promote them. This sudden interest was accompanied by a plundering of the last available family heirlooms by dealers. Fake ancient headscarves and shawls from the region of Imi-n'tlatet (in the southwest of Morocco) also appeared around this period.[1] They were easy to produce, as they only needed to be decorated with henna patterns. Old textiles, or brand new ones that had been yellowed, torn here and there, and submitted to various treatments in order to create holes, were used. Only experts could spot that the patterns were invented, or foreign to the region where they were supposed to come from.

The distressing of carpets occurs elsewhere in Morocco, but the practice is most intensively performed in the Sirwa, because it is the main source of carpets for the international market and other Moroccan tourist towns. Distressing is not performed with the sole view of selling to a small niche of collectors, but is done on a large scale to cater for a wide range of buyers. Dealers collect information about their buyers' taste by observing their reactions to the carpets they show them, and by reading collectors' books and exhibition catalogues on Moroccan weavings. This information is fed back to the producers. The practice of distressing carpets follows the same path: it is the obvious taste of tourists for 'old and dirty' things, as many locals have jested, that led Marrakesh dealers to subcontract the ageing of carpets to the Sirwa region, where labour is cheap and the activity can remain hidden.

The scale of the carpet distressing practice in the Sirwa became obvious during my ethnographic fieldwork among weavers there in 2003–4. I once came across a strange sight at the bend of a mountain: fields of carpets covering the ground of a village (Figure 15). I soon found out that in this village several middlemen worked for Marrakesh dealers, providing them not only with carpets but also with the service of washing, finishing and ageing weavings.[2] This was done on a larger and much more organized scale than elsewhere in the region, where similar work was occasionally carried out in the confines of the home. Here the villagers had built a concreted area around the main village fountain. Each middleman had his own designated area. Carpets might only be washed in water and detergent, to rid them of dirt and soften them, or might be subjected to more invasive ageing techniques such as burning and sun exposure to give them an antique look. These techniques allow middlemen to provide buyers with the products they seek at a lesser cost. Natural dyes have become very expensive, and apart from madder and henna are no longer readily available in Morocco. Finer threads are no longer spun as they require more work.

How to make an old carpet

When a newly woven carpet is finished in the Sirwa, it is first laid on a bare concrete floor and burned with a blowtorch on both sides. The aim of this operation is to get rid of the nap normally present in handspun home-produced wool. Burning also eliminates undesirable materials present

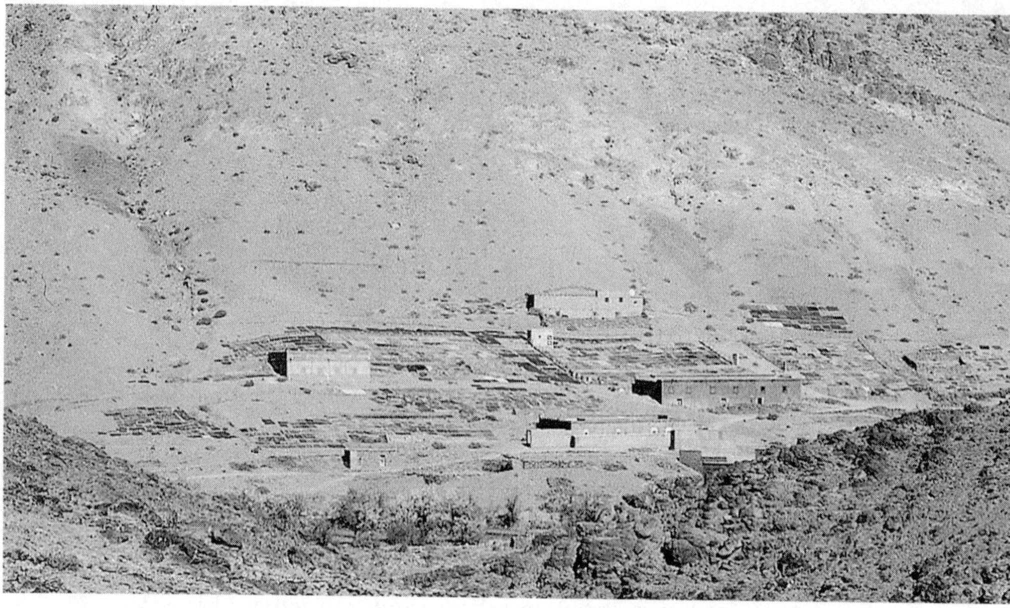

15 'Fields' of carpets surrounding a village where middlemen specialize in the washing and distressing of carpets

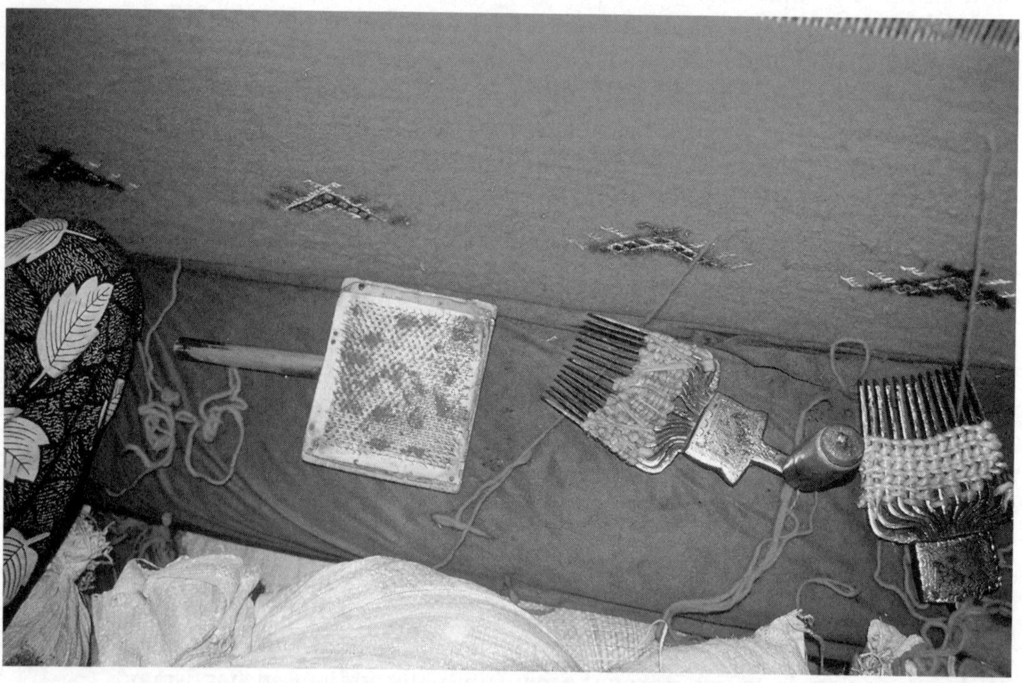

16 A card (centre) used by weavers to create a nap on a brocaded carpet

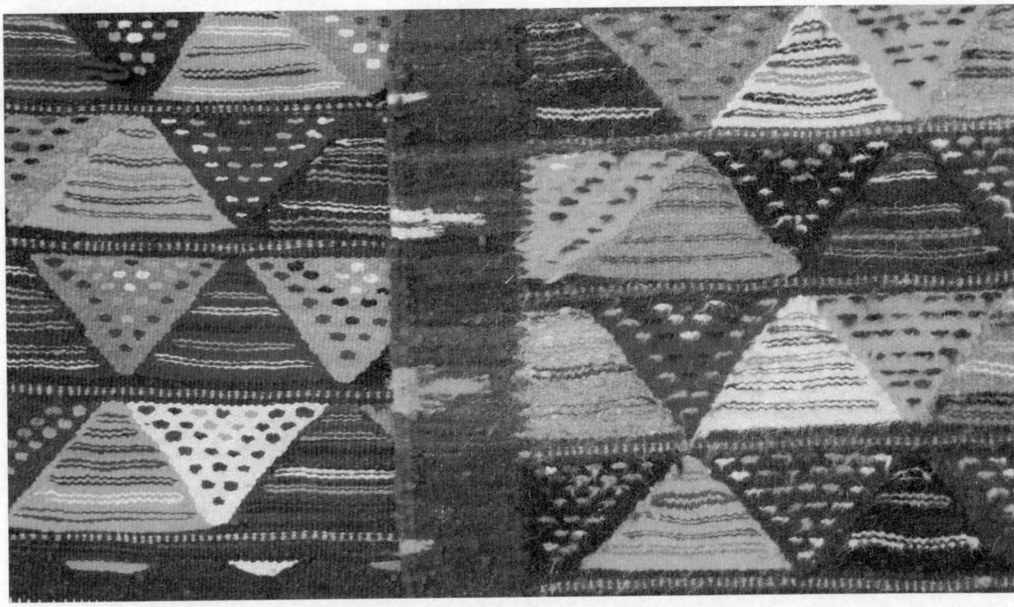

17 Detail of a tapestry woven carpet: on the right the untouched original weaving; on the left is the surface once it has been burnt, scraped and washed

in low-quality industrially spun wool. In spinning mills that produce the lowest quality wool, like the ones I visited in Rabat and Salé, the raw wool is not thoroughly washed and cleaned, in order to keep the weight of the yarn higher. Dealers encourage weavers who use this type of wool to create a nap by brushing the surface of the weaving as they make it with an old card (a flat wooden tool that has fine nails embedded in it). The napping operation raises the fibres at the surface of the carpet, slightly covering the brocaded motifs (Figure 16). Brushing the surface of the carpet pulls out fibres and dirt, enabling their easy elimination by carbonization. Finally, burning allows the practitioner to remove any technical faults protruding at the surface of the weaving, such as knots resulting from mending a broken warp thread. In the new type of carpets produced today, the intricacy of tapestry patterns involves numerous changes of colours, or 'stops'. The weavers cut the thread endings either with a razor or with their own fingers. Despite this, the endings appear on the surface and must be burnt away (Figure 17). The burning of the back or reverse side of the weaving removes the long threads necessary to decorate the front in the brocading technique (Figure 18).

Once the carpet has been subjected to burning, the remains of the carbonization as well as the smell must be removed. At this point the surface will have a carbonized and oily look, due to the presence of oil (lanolin) in the wool, but this is soon removed with a brush, falling into black powder

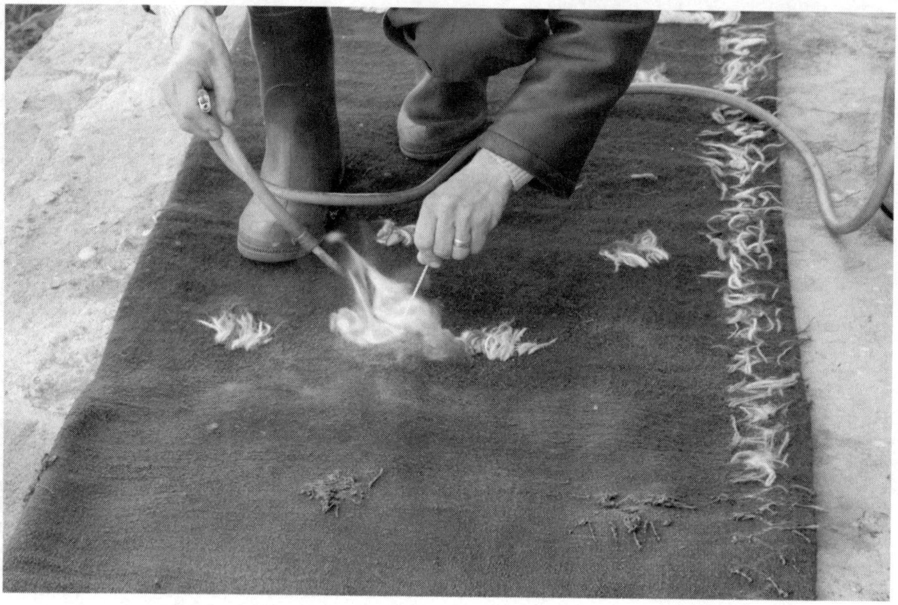

18 The surface of the brocaded black carpet is burnt with a blowtorch

on the floor. The carpet is then moved to a wet area on a gentle slope, where it is laid flat and covered with a small amount of washing powder and bleach that is mixed in a bucket of water. It is brushed in the direction of the weave (towards the fringes) with a hard bristle, plastic floor brush. Bleach removes the smell of the burnt protein. This also prevents the bleeding of colours and removes any excess dye. The carpet is then sprayed with clean water using a hose to remove the remaining washing powder and bleach. A scraper with an iron blade is used to push the dirty water out of the carpet, scrubbing in both horizontal and vertical directions and on both sides (Figure 19). The weaving is washed a second time to remove the chemicals from the fibre, the practitioner folding it as he presses the water out of the carpet. Finally he rolls it into a tube and lets it stand against a wall to allow the remaining water to come down. A few hours suffice for the carpet to dry in the harsh sun of this region. Exposing it longer will result in a slight fading of the colours. To tone it down further the exposure time needs to be over ten days. For a better effect, both sides should be 'tanned' otherwise the ageing work will appear more clearly. Sun exposure is also used to remove the bleeding of colours on white areas.

From historical documents and interviews with dealers and 'finishers' in the Sirwa, I believe that these distressing practices started in the late 1980s with the increased commoditization of carpets and the intensification of

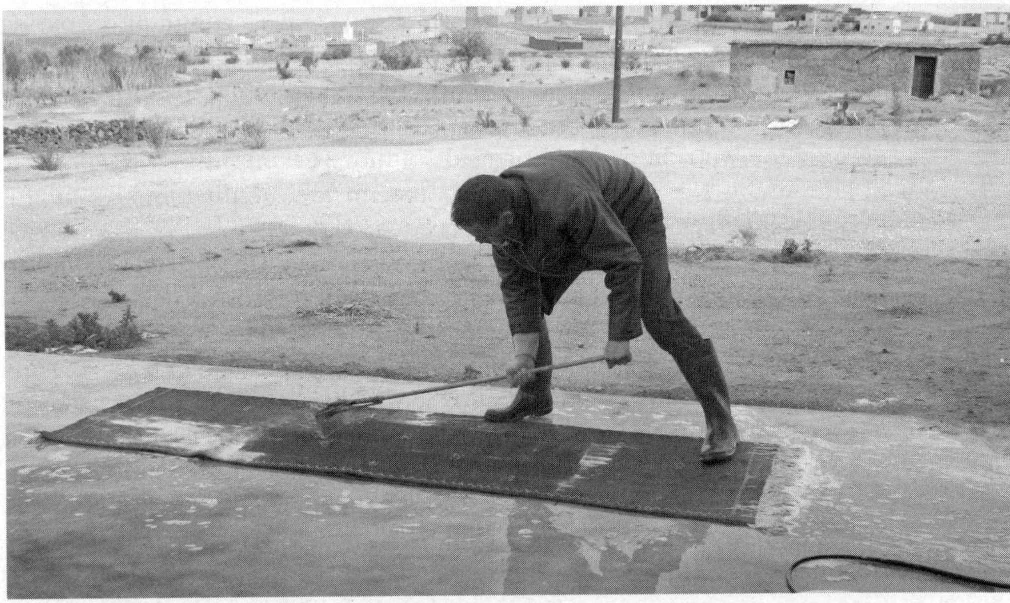

19 The carpet's surface is scraped both horizontally and vertically with a heavy-duty iron scraper

production, which accompanied the dwindling supply of older pieces. Where did they come from? This is not a simple matter, because some of the processes are extremely ancient and others relatively new. The process of raising a nap in order to crop it with shears or scissors was well established at the time of the Roman Empire, and it is likely that it was also known in Morocco, at least in the coastal cities. The colonial literature reports such methods being performed in Tunisia on flatweave textiles for the local market.[3] There is no doubt that the operations of burning and toning down carpets in Morocco were borrowed from oriental carpet producers who still use such practices (some even advertise the fact on the Internet). In fact, in the late 1920s, the head of the colonial Department of Craft in Morocco, Prosper Ricard, surveyed the situation of the oriental carpet industry in several countries in the Near East with the view to market and develop the production of Moroccan carpets as a source of revenue for the country. In Greece, he observed that artificially dyed oriental carpets were toned down using chlorine. Sometimes, motifs had to be touched up by hand-painting shading into the pile. The surface was brushed and scraped to remove any pile that was loose and after washing, cropped again to ensure a very even and thin pile. The back of the knotted carpets was blowtorched to obtain a similar clean and neat finish. Finally the pile was exposed to a hot iron to recover its lustre.[4] Yet these operations might also remind us of the finishing practices used in the industrial treatment of wool. For example, scouring traditionally involves simply washing the wool in water to remove vegetal and other unwanted matter before carding and spinning. In the industrial sector, scouring is now done through chemical carbonization. This involves dipping the wool in a sulphuric acid solution, heating it to a temperature of around 95–110 degrees centigrade, and then vacuuming and washing it to remove the carbonized vegetable residue.

Carpet distressing simulates the natural wear of the surface due to mechanical abrasion and photo-oxidization, but it is also a relatively cheap way of compensating for the lack of skills in dyeing and for the bad quality yarns and dyes. Distressing enables less skilled dyers to obtain pastel colours or colour ranges outside those readily available on the market by exposing textiles to the sun until they are toned down. Good wool does not need bleach to shine. And good quality dyes, fixed properly on the fibre and well washed, do not normally need to be eliminated after weaving.

In addition, these treatments compromise the object more substantially than it would at first appear. Both natural and synthetic dyes will fade with time, and the fuzzy surface of a new carpet will eventually become neat, but the surface of these carpets is abraded and corroded by mechanical and chemical means, accelerating the natural progress of time. These finishing methods not only give the carpet the appearance of an old artefact but also affect the integrity of the weaving. The burning and scrubbing of the surface removes the first layer of protection of the yarn. As with human skin, the photochemical degradation due to light exposure is one of the most important factors in the ageing of textiles. Bleaching and extended sun exposure weaken the fibre's elasticity and capacity to stand wear, degrading the weave. Consumers may get a textile they see as ready to be displayed, yet because of this accelerated ageing treatment, it will lose its colours faster. Like fast food, these are 'fast-carpets' that are made to give immediate satisfaction to the buyer.

Local conservation practices and beliefs

These distressing practices, which can be interpreted as the shortening of life of textiles, are directly opposed to local craft traditions and cultural beliefs. Wool in Morocco, like wheat and water, is seen as precious living matter, a gift from God to cover one's body and keep it warm and safe from illness. Before the intensification of production and the advent of machine spinning, in a context where wool was scarce and valued, weavings were made to last a long time. Sirwa people have tried to preserve and extend the durability of their textiles through various weaving techniques and conservation methods. Weavers developed special selvages, and used a twinning technique at the beginning and end of the weaving, to reinforce the textile's more fragile extremities. In addition, instead of covering the background of the knotted carpet with a single colour (almost invisible under the pile), they used to weave weft bands of different colours, so that when the screen of knots had completely worn out, probably after several generations, it would turn into a flatweave textile made of several coloured bands. Once the carpet enters a home, its owners minimize the risk of wear by going barefoot; when exposing carpets to the sun (as a way of getting rid of pests), they keep the colours as vivid as possible by exposing only the wrong side to the sun, and for a short amount of time. In families with means, heirloom carpets are folded carefully and stacked above each other, opened only when there are guests. In urban areas and more humid regions where moths are more active, lavender and mothballs are used as an additional protection. In the past, white weavings were exposed to sulphur smoke to revive the whiteness of the wool, which tends to take on more creamy tones with time. Thus wool is seen as a mutable matter, which needs to be revived, maintained and looked after. All these practices also can be seen as tampering with time, but in the opposite direction from distressing: here temporal progress is slowed by imparting surfaces with stamina. In this respect the attitude of a traditional Moroccan household to their carpets is not unlike the practice of collectors and museum staff.

Lessons about distressing

Thus informed, I set off to document the distressing operations with my own eyes. The prospect of burning a carpet did not enchant me, and so I purchased for the purpose of experiment a carpet whose colour and aspect did not appeal to me. It was a cheap flatweave black carpet with a few brocaded colourful motifs. It was sloppy and hairy, and its background colour reminded me of deep black artificial dye on white human hair. I cannot claim that I approached the encounter with the distressing practitioners objectively. I already had a negative prejudice about their work, and this was further grounded in my close contact with weavers and my knowledge of their work, energy and skill. I saw the work of these finishers as destroying hard labour and creativity. This was further reinforced when a dealer showed me a collection of older pile carpets that had been artificially faded to the extent that the composition and motifs disappeared almost completely into a whitish blend. They smelled of bleach and there was an electric static charge each time they were folded and unfolded, which indicated an excessive use of pile brushing.

And so I was not prepared to accept certain facts: not only that the work of these practitioners is a skill in its own right, which involves careful chemical and mechanical alteration of the surface; but also that a weaving's shortcomings could actually be improved by this treatment. The effect resulting from distressing my carpet was that it acquired a neat and even face weft texture, simulating that of natural wear. The colour of its surface now showed a range of browns, naturally found on the fleece of darker sheep, which had not been evident before. In addition, with the removal of the small wool fibres, the motifs appeared neater, and the weaving looked cleaner and finer. The surface had a more homogeneous texture. Thus these methods did bring improvement to the weaving, making it 'clear or free of that which is unwanted, i.e. impurities, adulterations, imperfections, blemishes,' to cite the theorist of surfaces Thomas Leddy.[5] The textile had been adulterated through the use of bleach; however, taking away its surface hair and other foreign elements gave it a smoother feel and look; fingers and eyes alike could slide across its surface without meeting any interference or obstacle. And when I brought it home, my house would not be invaded by loose fluff and sheddings from subsequent wear. This treatment provides the buyers with the type of object they want, a ready-to-buy-and-use carpet, in a society no longer concerned in investing in a sturdy furnishing made to last several generations; this is the right object to fit the aesthetic of their house, right now.

Local attitudes to distressing

Weavers and dealers in the Sirwa remark again and again, with evident bewilderment, that westerners prefer faded carpets. In 1926 Ricard explained that the toning down of oriental carpets was performed to cater for consumers used to lighter colours. Indeed, consumers of oriental carpets were shocked by the colours of Moroccan carpets, which they considered too gaudy, when they visited Morocco in the 1990s.[6] What is at stake here is not just that faded colour denotes antiquity and that carpet dealers often sell faded carpets as old ones; it is also a matter of perception of what a carpet ought to look like. Harald Böhmer, an Austrian chemist who worked with village cooperatives and a university to reintroduce natural dyes in Turkey, has tried to teach the public about this misreading of carpet surfaces, explaining that the naturally dyed carpets made today in Turkish cooperatives may have bolder and deeper colours, but that they will take on a beautiful patina over time and will be enjoyable for several generations.[7] Although some customers buy Moroccan carpets because of their bold and vivid, sometimes gaudy colours, which can seem compatible with a certain modernist taste, most consumers are quite conservative in their appreciation of these carpets. This is partly due to an education of the eye based on long familiarity with the aesthetics of carpets conserved in collections and museums. Arguably, as noted by Susan Crane, such artefacts are denied their intended lifespan;

they are frozen in a moment which fixes them as representative of a period or a category.[8] In other words, the carpets that we have come to admire in collections are representative of the arrested time when they were preserved. We do not see what they were like when they came off the loom. Through such experiences, anchored in a well-meaning art education, these buyers have learnt a normative way of appreciating non-western carpets and arts. Steiner has remarked that in their consumption of African artworks, tourists are looking not for the new but for the obvious and familiar, a pattern of consumption that is part of the more general structuring of the tourist experience. Christopher Steiner links this preference for the familiar to the western aesthetic experience of and taste for redundancy, the 'pleasurable repetition of an already known pattern'.[9] Maybe there is also some elitism here, or conversely a lack of confidence in one's own taste. Harry Gene Blocker, a collector of African artefacts recounting some of the problems of establishing authenticity, concludes that at the end what really matters is aesthetic sensibility: 'genuine pieces almost always have a recognizable air of superior quality about them that not even the best fakes can capture.' Blocker says that a good question to ask, confronted with a piece of doubtful age or provenance, is, 'could anyone today do anything that good?'[10] And that if the answer is no, one should buy it, even if it is not all that old.

Yet there is a contradiction in consumers' desire to possess an object that has patina, authenticity and perhaps a personal history, but has not acquired the other signs of age. Thus a carpet has to be neat, clean, evenly finished and have faded colours, but not dirt, food remains, holes or fleas. Carpets too worn out or too shabby – or in such a good state that their colours are still vivid – would not appeal to them, however authentic.

Ironically, for those seeking antique carpets, for whom time enhances the value of things, the providers of prematurely aged textiles perform the opposite of what the consumer of antiques seeks: real time inscribed in the artefact. There is no doubt that most consumers would be horrified to find out that their newly purchased antique is a commercial imitation of the natural decay of wool, the extreme opposite of authenticity: they are brand-new products for the western market, probably never used, intentionally distressed and submitted to extreme abuse. They think of themselves as appreciating the weaving skills inscribed in the carpets, which have actually been obscured, but are blind to the skills of falsifying temporality, the skills they are actually seeing.

Conclusion: local aesthetics and change

Today in the Sirwa, with the entry of carpets into the global economy, long-held beliefs around wool and weaving are dying out. The new pace of work and the introduction of new technologies and materials have led to the demise of rituals considered no longer necessary. Taste and fashion are ever changing and weavers are keen to keep up with them. Like their European counterparts in the 1950s–60s who preferred new plastic tables rather than oak ones, Moroccan consumers (urban and rural) have entered a phase in which material culture is ephemeral. So for Moroccan consumers, a carpet must be new. The middle-class housewife who can afford it will change the bedding and flooring of her living room at least once every ten years. Sofas are usually ordered from an upholsterer, who works with fabric brought by his client and adds the finishing (trimming, braids). The colour and aesthetic of sofas and carpet have to match. The same value is given to brand-new carpets in rural production areas where, even in well-off families, an opportunity to sell an old carpet at any price above the cost of the raw material is rarely declined. In fact, in families who produce regularly for the market, the production of pile carpets once or twice a year for their own household is often done with the knowledge that within five to ten years of use, they will probably be bought by a dealer. Consequently, in most houses in the Sirwa, a traditional centre of weaving, it is rare to find a carpet more than twenty years old. The availability and relative cheapness of the material and the frequency of production has transformed carpets into investments, family furnishings which are recyclable into cash.

Yet weavers' aesthetic sensibilities differ greatly from those of their buyers. Where some westerners see a modernist work, weavers see a sloppy, careless and awkward job, probably made by a beginner. Whereas international buyers are fooled by the surface quality of weavings, blind to the concealments and adulterations they have been subjected to, weavers believe that the surface should be scrutinized for the care, technical skills and creativity of the maker, which appear in the very structure and surface of the weaving. A beautiful carpet is one that is tightly woven and well beaten, one that shows the capacity of the weavers to align motifs in an orderly manner, a symmetrical arrangement that demands mathematical and compositional skills. Faded carpets are seen as dull and drab, lacking brightness and contrast; in their eyes only impoverished people buy old or aged objects. Weavers scorn the taste of Germans for natural colours (they compare their taste to that of their grandmothers who did not have access to the same range of colours). They enjoy deep hues like dark blue and vivid red, rather than pale or mid-tone colours such as light blue or green and orange.

Distressing methods contribute to the falsification of temporality, inscribing patina into the surface of new textiles and imposing upon them a visual and tactile order. This set of processes conflicts with the aesthetic sought locally in the Sirwa, which is based on the sharp and symmetrical contrast of colours and shapes. Yet an approach that understands distressing only as the accelerated erosion of the surface, or which focuses only on the shortened lifespan of these distressed textiles, ignores the efficacy of the distressing process. In a context where buyers seek to buy cheap, albeit ephemeral products, it is a way of improving the finish of textiles made of low-quality materials. Dealers give carpets the aesthetic look that their consumers expect, contributing to the symbolic and material transformation of carpets into valuable and desirable objects. As long as this process is understood by the buyer who purchases the carpet, there is nothing wrong with this practice. Although distressing methods are to some extent opposed to local practices and values, they have also become part of the process of creating value through textiles – allowing these weaving communities to generate an income, and to carry on their trade into the future.

Notes

1 M.-R. Rabaté and F. Sorber, *Costumes Berbères du Maroc: décors traditionnels / Berber Costumes of Morocco: Traditional Patterns* (Paris: Éditions ACR, 2007); M.-R. Rabaté, 'Vrai Costume, faux tissus: les méfaits de la gloire', *Qantara*, 66 (2008), 44–46.

2 The most successful middlemen belong to the Ayt Waghrda ethnic group, famous for the skills of its weavers.

3 L. Poinssot, *Tissus décorés de Gafsa et imitations* (Paris: Horizons de France, 1953), pp. 44–46.

4 P. Ricard, 'Notes au sujet de la fabrication des tapis au Proche-Orient', *Hesperis*, 6 (1926), 435–436.

5 T. Leddy, 'Everyday surface aesthetic qualities: "neat", "messy", "clean", "dirty"', *Journal of Aesthetics and Art Criticism*, 53:3 (Summer 1995), 265.

6 N. Purdon, 'Black tents and palaces', *Hali*, 17 (1995), 82–91.

7 See J. Anderson, *Return to Tradition: The Revitalization of Turkish Village Carpets* (San Francisco: California Academy of Sciences/Seattle: University of Washington Press, 1998).

8 S. A. Crane, 'The conundrum of ephemerality: time, memory and museums', in S. Macdonald (ed.), *A Companion to Museum Studies* (Malden, Massachusetts: Blackwell, 2006), pp. 98–109.

9 U. Eco, 'Innovation and repetition: between modern and post-modern aesthetics', *Daedalus*, 114:4 (1985), 161–184, quoted in C. Steiner, 'Authenticity, repetition, and the aesthetics of seriality: the work of tourist art in the age of mechanical reproduction', in R. Phillips and C. Steiner (eds), *Unpacking Culture: Art and Commodity in Colonial and Postcolonial Worlds* (Berkeley: University of California Press, 1999), pp. 87–103.

10 H. G. Blocker, *The Aesthetics of Primitive Art* (Lanham, Maryland and London: University Press of America, 1994), p. 186.

6 ✧ Textiles and time: conservation and public response

Margaret Ponsonby

HISTORIC interiors are a combination of structure and contents: the architectural shell, the interior decoration of the walls and the many items that constitute the furnishings. It is the fragile interior surfaces upon which the passage of time particularly leaves its mark. The effects of the ageing process will continue even if they are preserved in ideal conditions, and historic houses can never provide those. Preserving and keeping on show all the historic elements retains the authenticity of the interior but this process is far from straightforward. How preservation should be dealt with has changed according to the methods available at any time, and according to the desired outcome. It is the evolving nature of conservation – the physical result of different kinds of intervention in the conservation process and how this has been shaped by cultural attitudes to what constitutes a historic interior – that is the focus of this chapter.

Decisions have to be made regarding the treatment of the various elements within the historic interior, particularly the treatment of the most fragile and fugitive elements: the textiles, wallpaper and paintwork. All these aspects of historic interiors are vulnerable to time and decay but also to loss through replacement. Textiles are probably the most delicate features of any interior. They suffer from exposure to light, damp and insect infestation. Opening a house to the public adds to the problems with an increase in dust circulation and exposure to accidental and deliberate contact by visitors. Textiles become faded and discoloured. Fibres break down, shatter, and eventually disintegrate. Various methods can be used to prevent such damage using blinds and shutters at windows and UV filters attached to the glass. Opening hours can be limited and the rest of the time the rooms can be left in darkness.[1] However, nothing can completely protect old textiles exposed to the conditions that prevail in historic houses. Similarly, wallpaper will fade and react to damp conditions. Historic paint changes due to chemical reactions over time. The build-up of paint layers blurs the outlines of mouldings but if this is removed then the historical record of earlier decorative schemes is lost.[2]

The conservation profession depends increasingly on scientific methods and now favours minimal intervention in order that the historical document remains intact. However, the decision-making is rarely clear-cut and a variety of options is usually open to a conservator treating an object or a decorative scheme. The methods chosen, while aiming to protect the integrity of the historic surface, are also influenced by the interpretive method employed at a particular site, which in turn is influenced by who owns the house.[3]

The method chosen will influence the nature of the historic interior. In the case of textiles, modern techniques use conservation net that has been dyed to a similar colour and this is stitched in place to hold all the loose fibres down and prevent further disintegration. This net has the effect of softening outlines and adding to the faded appearance of furnishings. The decision to keep old textiles in a historic house affects not only those individual objects but creates a particular ambience in the rooms – they will be faded and patinated. In addition, lighting, both natural and artificial, will need to be kept to a minimum in rooms with such objects. The retention of objects with patinated and shabby surface qualities, sometimes with the additional surface layer of conservation, gives a sense of authenticity but it also produces an obscured view of how the rooms would have looked in the past. This, together with the low lighting, influences the experience for visitors.

A different method of protecting old textiles is to retire them – that is, put them into storage and put modern replicas in their place. The decision to retain or to replace textiles makes a major impact on a historic interior. Replacement alters the 'tone' of a room and the relationship between all the objects it contains. The treatment of the textiles can also tip the balance between a preserved interior and one that is restored to an earlier period: if the textiles are replaced then redecoration using period colours, which will look much brighter than the faded colours of aged paintwork and wallpaper, is more likely to take place. In addition, the lighting in these rooms does not need to be reduced to protect the furnishings. Houses that have been given this treatment present interiors that are closer to how the original inhabitants would have experienced them. The subtle differences in the surface qualities of new materials in the form of textiles, paint and wallpaper thus produce a marked difference in the interpretation and alter the experience for visitors.

Within the profession of textile conservation there has been general agreement concerning aims and outcomes in Western Europe and North America. The notion of preventive conservation is widely accepted, as is the desire to use the least invasive methods. Despite the scientific nature of conservation and the rational decision-making that this might suggest, the concepts that govern the level of work to be carried out on an object are not only decided by physical considerations but are also culturally determined. Textile conservator Dinah Eastop has commented that the agreed principles are 'open to wide interpretation and [support] a very broad spectrum of justified practice'.[4] This chapter aims to explore how this 'broad spectrum' is influenced by different cultural attitudes in England and the USA and how this affects the treatment of historic interiors open to the public.

Interpreting historic interiors in England and the USA[5]

It is important to note that there are broad differences in how historic interiors are treated in the two countries. Many houses in England display 'history in continuum', that is layers of occupation that have accumulated over time, whereas in the US it is more usual to choose a 'period of significance' and to base the interpretation of the house on that. This difference of approach came about when historic house museums first emerged in both countries in the mid-nineteenth century. In America, the original impetus for preserving houses and other historic sites was patriotism and this continued into the twentieth century with many houses, such as Mount Vernon, Washington, DC, having associations with presidents and other important figures in American history.[6] Various female organizations, such as the Daughters of the American Revolution and the Colonial Dames, and individual historical societies were the prime movers and continue to be the owners of the majority of historic house museums. By contrast, in England there had long been a tradition of privately owned homes opening to the public, if a rather select one.[7] When local councils and charities, including the National Trust, became responsible for some historic houses, they were continuing this tradition but with wider participation. These differences have established subtly different priorities; in the US the emphasis is on the story that the building represents, whereas in England the emphasis is on the architecture and collection accumulated by the family over time.

The result is that in England houses are more likely to display faded and patinated interiors whereas in the US they are more likely to be restored to the original colours. Therefore, in England historic textiles tend to be conserved and retained in situ whereas in the US it is commonly felt desirable to retire historic textiles and replace them with replicas.[8] To illustrate how the varying interpretive aims have shaped historic interiors, two examples will be examined, one in England and one in the US. While using different styles of preservation and interpretation, both also display how methods have changed over time. The result is rather different interiors and different kinds of history for public consumption.

An English country house: Uppark, Sussex

Uppark was built in c. 1690 and purchased in 1747 by Sir Matthew Fetherstonhaugh. Sir Matthew and his wife Sarah acquired a fine collection of paintings and objects for the house on their grand tour and commissioned many further items in England.[9] They also made alterations and carried out redecoration of various rooms up to Sir Matthew's death in 1774 when his son inherited. Sir Harry also added to the contents during his grand tour and commissioned Humphrey Repton to make some alterations between 1811 and 1815. Sir Harry finally married in 1825. His surprising choice of a wife was his dairy-maid who was fifty years his junior. Mary inherited Uppark when Sir Harry died in 1846 and she and her sister Frances continued to live at Uppark for many years making few changes and indeed with the intent of preserving the house and its contents as they had been in Sir Harry's time. Uppark passed to Sir Herbert Meade in 1930. He added the Fetherstonhaugh name to his own and he and Lady Meade-Fetherstonhaugh continued the tradition of caring for and preserving the house and its interiors. Among the collections of paintings and furniture there were also a great many textile items such as carpets, bed hangings and curtains. All the original colour schemes were retained. Uppark became known as a house that had been unaltered, a 'sleeping beauty' that was only disturbed when the house was devastated by fire in 1989.

The continuity of the furnishings up to this time was greatly helped in the twentieth century, by the conservation work done by Lady Meade on the textiles of window curtains and bed hangings.[10] This ensured their survival and in turn contributed to the shabby grandeur that the house was noted for (Plate 12). Likewise, when the National Trust took over in 1954 they did so with the intention of changing as little as possible. The Trust used John Fowler as a consultant on the interiors, but he advised keeping the decorative schemes intact with touching up of paintwork where necessary and washing down some rooms. At other houses Fowler had imposed his own taste that either disregarded the historical evidence or recreated schemes based on inaccurate methods such as paint scrapes. This method quite literally scrapes away the paint to reveal earlier layers but this cannot take into account how the ingredients in the paint change over time.[11]

In the 1950s and 1960s while a more scholarly approach was gradually being adopted there was not the desire for historical accuracy that pertains today. By the 1970s, it was decided to freshen up the schemes somewhat, perhaps influenced by more thorough restoration schemes being carried out elsewhere at the time.[12] Once again, Fowler was consulted and on his advice, several rooms were repainted using the same colours. Despite the changes that were made the interiors presented uniformly faded paintwork, wallpapers and textiles. Everything in the interiors appeared to have aged together.

The fire of 1989 destroyed the upper part of the building. Although ninety-five per cent of the contents of the ground floor, where the main reception rooms are located, were saved, the walls and decoration were badly damaged.[13] After carefully considering all options, the National Trust took the decision to reinstate the building and its decoration to how it was the day before the fire, restoring all the layers of history that had then been present.

The enormous and complicated task of restoration resulted in numerous instances of preserving original materials and infilling with new work where necessary. The treatment of window curtains illustrates the Trust's philosophy. When these were conserved by Lady Meade in the 1930s her methods involved stitching parallel lines to incorporate all the loose threads. Such methods have now been superseded by the use of conservation net that holds loose fibres in place but requires far less stitching (closely worked tramlines do additional damage to shattered textiles). Despite the changes in procedure, the textile conservation at Uppark following the fire preserved her work and 'imitated [it] where broken by fire damage'.[14] This decision was taken as part of the general philosophy for dealing with the fire damage at Uppark, to return the house to its condition the day before the fire. An additional benefit was derived from including the story of the fire in the presentation. The conservation was meant to preserve the unaltered 'sleeping beauty' image of the house while being honest about the enormous change that it had suffered. One textile example of this is the window curtains in the saloon. The fire variously affected the four curtains, one of them being destroyed. Careful conservation was done on the three that had more or less survived intact but the fourth curtain was replaced with new fabric, in as near a match as possible. This fourth curtain, however, is clearly a brighter, cleaner textile. The colour could have been toned down to harmonize with the others, but it was decided to allow the discrepancy in colours to remain as evidence of the conservation work following the fire (Plate 13).[15]

How the restoration work will age is an unknown factor and it is acknowledged that the work 'inevitably bears the stamp of the period in which it is done'; this might result in a rethink of strategy in time.[16] The Trust undoubtedly reacts to visitor perceptions as well as adopting new preservation methods. However, in recent years the National Trust has invested time and money in educating the public on how best to preserve historic interiors using explanatory displays in many of its properties. This has been done at Tyntesfield where visitors are invited to 'adopt' objects that require conservation, and during the Attingham Re-discovered Project visitors were invited to comment on how the interiors should be treated during the conservation process.[17]

An American colonial mansion: the Moffatt-Ladd House

The Moffatt-Ladd House in Portsmouth, New Hampshire, was built in the early 1760s and lived in by various branches of the same wealthy merchant family until 1911.[18] It was furnished with locally made furniture but also with imported goods, damask hangings and curtains, gilt looking-glasses, Persian carpets and wallpapers. The house was offered to the Colonial Dames of New Hampshire in 1911. At first, the Society used the house as their headquarters but also as an historic house museum. The house was thought to be a particularly fine example of colonial architecture and it had been the home of one of the signatories of the Declaration of Independence, William Whipple.[19] The various branches of the family had preserved many items in the house including portraits, some furniture and curtains, and the layers of historic paintwork. However, when it was first opened to the public the house was not presented as a historically correct interior would be understood today. As Jane Nylander has observed, the house was originally displayed 'in a "tasteful" style that reflected more of the comfortable appearance of members' homes than it did serious curatorial research'.[20] Indeed, members donated items of furnishings for the house from their own homes. This was not an unusual situation and the prevalence of the Colonial Revival style in furnishings extended from private homes to most houses open to the public in the US at the time. Early advice at Moffatt-Ladd came from Luke Vincent Lockwood, curator at the Brooklyn Museum. He had seen the house previously and therefore felt able to make suggestions without a further visit. He advised that the floors should be stripped and waxed, and the woodwork stripped of all previous layers and painted a 'dead white'. The surviving blue and yellow curtains in the parlour suggested the colour scheme for the room since they were a 'pure Colonial combination'.[21] Blue and gold were also the Society's colours and this influenced the choice of wallpaper for the Long Gallery when it was redecorated in 1939. A design from 1815 was chosen partly because it was in blue and gold.

In 1936, two inventories of the house for 1768 and 1786 were found but had little effect on the decision-making. However, in the 1950s a new direction was taken in the interpretation of the house with 'a new emphasis on authenticity'.[22] A cut-off date of 1825 was established for the restoration and for any new accessions for the house. The two inventories were to be used as part of a more scholarly approach to interpreting the rooms to different periods in the history of the house and its occupants. Paint scrapes were used to establish the sequence of paint colours. This resulted in the early 1970s in one bedroom being interpreted to the time of William Whipple's occupancy in 1785 with green paintwork, although using modern paints as they were found to be more stable. The bed hangings were made in reproduction textiles suitable for the period (Plate 14).

The evolving methods of interpretation reflect how Moffatt-Ladd House has employed 'the best practices of the time'.[23] By the 1990s, for example, paint scrapes were replaced with microscopic paint analysis.[24] The cut-off date was eliminated so that the history of the house and its occupants up to 1900 could be reflected. What had been used as the front parlour in the eighteenth century had become the dining room in the nineteenth and this had been recreated using the blue damask curtains that had survived intact in the house. These draperies had been used in the early days of the Colonial Dames' time at the house but have since been 'retired', with replica textiles hung in their place (Plate 15).

Experiencing the historic interior

The two houses, Uppark and Moffatt-Ladd, provide examples of the changing attitudes to how historic interiors should be conserved and interpreted over time and the different emphasis in the two countries. While the houses vary to an extent in size and grandeur the style of interpretation has a far greater influence on the experience for visitors. The Moffatt-Ladd House feels uniformly bright and clean despite the different periods on display and the different times that the work was carried out. The rooms are flooded with light without blinds at the windows and the reproduction carpets allow easy access to the rooms. Tour guides tell the history of the house and its occupants to groups of visitors.[25] The interiors are well researched and represent historical periods, although producing a replica of the past that appears authentic is always difficult.[26]

Uppark, by contrast, presents faded and patinated interiors to which have been added the conservation and restoration of objects and decorative shcemes following the 1989 fire.[27] The visitor 'reads' the interiors on a number of levels; the whole history of the house is present in every room. The interiors have a faded grandeur. The lighting levels are low, some historic carpets are rolled back to prevent contact with feet. The work of conservators contributes a further dulling of surfaces and sometimes a mottled effect of old material and the new infill. As in most National Trust houses the usual method of accessing the house is free-flow visitation with room stewards to answer questions.

In 2008, the education officer for the American organization, the National Trust for Historic Preservation (NTHP), visited Uppark and wrote up his reactions in a blog. He confessed he was mystified by the decisions taken by the National Trust. He could not understand the philosophy of reinstating the house to its condition the day before the fire since 1989 was not 'particularly significant' and, therefore, surely it would have been 'wiser to evaluate the significance of the site and then restore the building to that period'.[28] This view misses the point that if the interior decorations had been restored to their eighteenth-century splendour, then every single item in the house would have looked out of place. To the employee of the NTHP the reasoning behind the conservation of Uppark was wrong but also he had been conditioned to expect interiors to be restored to a period of significance and, therefore, to look newly decorated. While some UK properties have been restored to a particular period, they remain a minority and most houses, particularly where the collection remains intact, prefer to follow the history in continuum method of interpretation. Visitors in England are conditioned to expect houses to reflect how the house has become over time rather than how it once looked.

Preserved as found interiors

The treatment of Uppark following the fire built on methods that the National Trust had been following for some years. The idea of preserving intact the interiors, furniture and furnishings in an arrested state of decay began with Erddig in Clywd and Calke Abbey in Derbyshire. English Heritage used a similar treatment at Brodsworth Hall in Yorkshire. The National Trust has continued to use this method at Tyntesfield, Somerset and Sunnycroft, Shropshire.

However, this method has not been used much in the US. Historic New England, perhaps the closest American organization to the English National Trust, has preserved the interiors of some of its properties such as the Codman House in Massachusetts.[29] Drayton Hall, South Carolina, a NTHP house has its original paintwork conserved to preserve the faded and patchy surface intact but the house is presented unfurnished.[30]

The idea behind preserved as found interiors is that when the property becomes a museum, the interiors will remain as they were when the house was lived in with all the contents remaining in situ rather than some things edited out. Any conservation work undertaken would preserve the condition of an object rather than restore it. Everything is valued despite its appearance. In particular, minor items recall the daily lives of the inhabitants as well as revealing the clutter that many people live with.

To the casual visitor these houses present an impression of being quite literally preserved in the condition they were in when 'found'. However, a long period of conservation work is always required and some changes are inevitable. Furniture needs to be re-arranged to allow the house to function as a museum. Some items are also replaced, in particular textiles, since their condition can sometimes be too fragile to continue in use. The changes that have been wrought are subtle and not easily detected.[31] While the historian might find this frustrating and want more information, for many visitors there is pleasure to be gained from such houses. The very vagueness regarding 'period' is an asset and allows visitors to identify with the interiors. This is due to the use of objects with patination, that are sometimes shabby and often of poor quality. Such ordinary items are often the most appealing and meaningful to visitors. The jumbling together of items from different dates also reflects our method of storage and retrieval from memory rather than the past being linear and in strict sequence.[32]

Creating patination through 'imported' objects

While such houses as Calke Abbey and Sunnycroft provide ambience through old and patinated objects that are related to the house and its former occupants, this is not usually possible when more humble homes are opened to the public. It is rare for a working-class home to survive with the contents intact, so non-original or 'imported' objects have to be used. The National Trust's Back-to-Back houses in Birmingham, West Midlands, and the NTHP's East Side Tenement Museum in New York city are similar projects.

Both of these sites have been restored to represent a number of different periods to reflect the lives of the inhabitants of the property during the nineteenth and twentieth centuries. The homes are furnished with period objects that are, overall, unconnected with the original inhabitants.[33] These objects are of low monetary value and are from the relatively recent past so that sufficient examples exist for the replacement of those used if need be. Therefore, the use of period objects does not prevent use of the interiors for educational purposes since the contents can be sacrificed to the wear and tear of small rooms where objects are exposed to visitors. However, these interiors do not replicate how they were when in use at the given time periods. When these homes were lived in, some of the items would have been new if of poor quality, so colours are duller and objects are more tarnished than they may have been at the time they represent. This perhaps establishes the general ambience of the interiors as impoverished although they may originally have contained bright and gaudy colours. It is not surprising this is done in England given the prevalence of preserved interiors. It is, however, a curious decision for the NTHP to have made and one that creates a particular message about this property in comparison with others of similar status but in rural locations.

Generally, living history heritage sites in the US recreate the colonial and early nineteenth-century periods. Properties such as Strawbery Banke, New Hampshire and Sturbridge Village, Massachusetts, contain modest rural interiors that are created using replica items. In particular, replacement textiles are always used. These replicated interiors suggest perhaps a wholesome and cleaner way of life in comparison with the Tenement Museum, which has an active educational programme that addresses difficult issues around immigration and poor housing conditions in New York city today.

Conclusion

What is the result of these decisions? As the last examples show, a completely different story is told through the presentation of the historic interiors. The retention of textiles with aged patinated surfaces affects the authenticity of individual objects but also creates the look and feel of an interior giving a sense of a profound history made up of layers of the past. Houses that are restored to a period using pristine fresh textiles can offer a valid and useful interpretive tool particularly where the collection has been lost and it is the only sensible thing to do. However, this method risks putting the chosen narrative so much in the foreground that the visitor assumes a passive role and loses the value of seeing authentic objects and interiors. Adopting a lighter touch that allows the surface of an interior to remain a 'faded canvas'[34] perhaps increases the possibilities for imaginative interaction and that is something which historic houses are particularly well suited to provide their visitors.

Acknowledgements

I began this research in 2004 whilst on Arts and Humanities Research Council (AHRC) Study Leave and with a Research Fellowship from Winterthur, Delaware. In 2008, I did additional research with the support of a grant from the Design History Society. I am grateful to all three funding bodies for their support.

Notes

1 See F. Boersma (ed.), *Unravelling Textiles: A Handbook for the Preservation of Textile Collections* (London: Archetype, 2007); *National Trust Manual of Housekeeping: The Care of Collections in Historic Houses Open to the Public* (London: Butterworth-Heinemann, 2005).

2 See English Heritage, *Layers of Understanding: Setting the Standards for Architectural Paint Research* (Shaftesbury: Donhead, 2002).

3 In England a great many houses are still owned and lived in by private families; many are members of the Historic Houses Association. Private owners are under

no obligation to adhere to historical accuracy and to retain furnishings intact. This chapter focuses on English houses owned by the charity The National Trust for England and Wales.

4 D. Eastop, 'Conservation as material culture', in Christopher Tilley, S. Kuechler, M. Rowlands and P. Spyer (eds), *Handbook of Material Culture* (London: Sage, 2006), p. 518.

5 This chapter is confined to examples from England rather than from all parts of Britain since different circumstances influenced the building of country houses in the past and to some extent heritage provision in the present. See note 3.

6 P. West, *Domesticating History: The Political Origins of America's House Museums* (Washington and London: Smithsonian Institution Press, 1999).

7 A. Tinniswood, *The Polite Tourist: A History of Country House Visiting* (London: The National Trust, 1998).

8 It is difficult to ascertain when the routine retiring of textiles began but a reference to the beginning of this move at Winterthur Museum, Delaware, in 1980, is recorded in M. Fikioris, 'Textile conservation for period room settings in museums and historic houses', in J. C. Adams (ed.), *Preservation of Paper and Textiles of Historic and Artistic Value II* (Washington, D.C: American Chemical Society, Advances in Chemistry Series 193, 1981), p. 274.

9 For the history of the house and the restoration following the fire see C. Rowell and J. M. Robinson, *Uppark Restored* (London: National Trust, 1996).

10 C. Rowell, 'Conservation traditions at Uppark: Lady Meade-Fetherstonhaugh as a textile conservator', in K. Marko (ed.), *Textiles in Trust*, Proceedings of the Symposium held at Blickling Hall September 1995 (London: Archetype, 1997), pp. 96–107.

11 For an account of John Fowler's work see H. Hughes (ed.), *John Fowler: The Invention of the Country-House Style* (Shaftesbury: Donhead, 2005).

12 Rowell and Robinson, *Uppark Restored*, p. 175.

13 *Ibid.*, p. 44.

14 Rowell, 'Conservation traditions', p. 105.

15 See K. Marko, 'Continuing conservation traditions at Uppark: the conservation of the saloon curtains', in Marko, *Textiles in Trust*, pp. 108–117.

16 Rowell and Robinson, *Uppark Restored*, p. 173.

17 S. Kay, 'Re-instating Regency décor', *Icon News*, 16 (May 2008), 4–5.

18 For the history of the house and its furnishings see B. McLean Ward, 'House and home: the Moffatts, Whipples, and Ladds, 1763–1911', in B. McLean Ward (ed.), *The Moffatt-Ladd House: From Mansion to Museum* (Portsmouth, New Hampshire: Moffatt-Ladd House and Garden, 2007), pp. 9–35.

19 For an account of the interpretation and restorations see J. C. Nylander, 'Museum and headquarters, 1912–2006', in McLean Ward, *The Moffatt-Ladd House*, pp. 36–55.

20 *Ibid.*, p. 39.

21 Quoted *ibid.*, pp. 38–39.

22 *Ibid.*, p. 44.

23 *Ibid.*, p. 51.

24 This method was first used at Mount Vernon. E. O. Welles, 'George Washington slept here ... and here's how it really looked: scientific paint analysis and Washington's

own writings help conservators give Mount Vernon a surprising new look', *Historic Preservation*, 37:5 (October 1985), 27–33.

25 The style of access influences the experience of the visitor. An organized guided tour will inevitably provide visitors with a prepared history of the house.

26 See M. Ponsonby, *Stories from Home: English Domestic Interiors, 1750–1850* (Aldershot: Ashgate, 2007), chapter 6.

27 Grant McCracken has argued that patina ceased to be a marker of status in the eighteenth century when fashionable consumption took its place. However, this argument does not take into account the rise in the same century of collecting antiques. By the time that houses were being preserved in the twentieth century, patina had become firmly associated with 'old' money in England. G. McCracken, *Culture and Consumption: New Approaches to the Symbolic Character of Consumer Goods and Activities* (Bloomington and Indianapolis: Indiana University Press, 1988). For the growth of the trade in antique furniture see C. Edwards and M. Ponsonby, 'Desirable commodity or practical necessity? The sale and consumption of second-hand furniture, 1750–1900', in D. Hussey and M. Ponsonby (eds), *Buying for the Home: Shopping for the Domestic from the Seventeenth Century to the Present* (Aldershot: Ashgate, 2008), pp. 117–138.

28 Max van Balgooy blog, http://historicsites.wordpress.com/2008/10/06/english-preservation-issues/, accessed 9 December 2008.

29 For a history of this organization see J. M. Lindgren, *Preserving New England* (New York and Oxford: Oxford University Press, 1995).

30 I was told by a curator at another NTHP site that visitors would think that collections, especially textiles, were being neglected if they were displayed in less than pristine condition.

31 For discussions around these issued with regard to Brodsworth see G. Chitty and D. Baker, *Managing Historic Sites and Buildings: Reconciling Presentation and Preservation* (London: Routledge, 1999); D. Philips, *Exhibiting Authenticity* (Manchester: Manchester University Press, 1997).

32 C. Saumarez Smith, 'Museums as Memory Bank', *Prospect* (July 1998), 80–81.

33 At the Birmingham Back-to-Backs the house that continued to be lived in by the tailor George Saunders does contain items from his home and business. Here facsimile items are used as a handling collection. Information from telephone conversation with Jeffrey Haworth of the National Trust, October 2008.

34 Chitty and Baker, *Managing Historic Sites*, p. 93.

7 ✧ Surface engagements in Astana

Victor Buchli

I would like to consider here the shifting material registers of buildings, to consider the vital importance their surfaces have for the constitution of social life. As in other arenas of modernist discourse, our understandings of architectural forms have been marked by a prejudice that privileges form over surface. Anthony Vidler notes the emergence of the 'épure' which mortifies the sensuous reality of architectural experience into a formal drawing, allowing it to be more easily communicated and stabilized across time and space.[1] The more recent conflation of drawing with built form heralded by new technologies such as CAD, laser cutting and 3-D printing amounts to a conflation of architectural form and its representation, notably at the expense of our understandings of the surfaces of these forms.[2] This is part of a continuing tendency within architecture to privilege the visual power of form, and the media used to stabilize these visualist registers, over other sensorial domains that characterize the architectural experience.[3] Following Mario Carpo one can chart a deeper history of this preoccupation with visual form, going back to the architectural representations emerging in the wake of the Renaissance and the need to visualize the notoriously imageless translations of Vitruvius that saw the development of drawings of the classical orders and their spread, which could only be facilitated by the emergent new technology of the printing press.[4]

It was, however, only recently that the modernist preoccupation with the primacy of form (couched in terms of function and honesty) prevailed over any other understanding of architectural elements. It is worth noting the nineteenth-century writings of Gottfried Semper, which rather than considering the primacy of forms and the outlines of buildings and walls, privileged surface, notably the decoration of walls:

> Hanging carpets remained the true walls, the visible boundaries of space. The often solid walls behind them were necessary for reasons that had nothing to do with the creation of space; they were needed for security, for supporting a load, for their permanence, and so on. Wherever the need for these secondary

functions did not arise, the carpets remained the original means separating space. Even where building solid walls became necessary, the latter were only the inner, invisible structure hidden behind the true and legitimate representatives of the wall, the colourful woven carpets.[5]

Semper argues for the primacy of the *Wandbereiter* (wall fitter) and of wall dressings, such as woven textiles, carpets, stucco, painted decoration, etc., over the walls themselves. He very usefully suggests a perspective that sees architectural form as anterior to the surface laid atop it, which is the primary site of social engagement and meaning. From this perspective, the wall is only the necessary support system for the surface, which is the 'true and legitimate' zone of architectural signification.

More recently in anthropology and other social sciences, surfaces have re-emerged as important arenas of engagement. A notable early work is Grant McCracken's thesis on 'homeyness' which contrasts the alleged sterility of modernist forms with those practices in the domestic realm which constitute moral forms of personhood associated with certain domestic arrangements. McCracken observes how the layering of fabrics, throws, patterns, drapes, lace, etc. envelops an individual, in a multi-sensory visual and tactile way. He argues that it is this surrounding that produces the moral terms of hetero-normative home life and sociality.[6] The material effects of layering are articulated in other terms by the anthropologist of art Alfred Gell. In his writing on 'tackiness', Gell argues that elaborately carved and painted decorative elements produce a sense of 'enchantment' that literally captures the observer, so that he or she becomes attached or engaged, or sticks, to the object.[7] The sense of revulsion that the *double entendre* of 'tackiness' suggests hints at the way in which this attachment and binding produces an embodied sense of disgust (or abjection, as Julia Kristeva would have it).This speaks to the violation of boundaries that may occur at a decorated surface, a conflation and oscillation of two things that should be otherwise held separate. Both McCracken and Gell, then, emphasize how elaborate surfaces entangle through their material properties (one through layering, and the other through adherence). In the case of McCracken the product is a certain moral disposition within social life; in the case of Gell, an implication of the body and its ontological attachments to material things.

This issue of detachment and separation has often been observed in the development of modernist sensibilities of moral personhood and architectural surfaces. In my earlier work on Soviet Constructivism I have argued that it was precisely this detachment from surface effects that was considered politically and socially productive. Whiteness, unadorned surfaces, polished metal surfaces and glass were all promoted to free the early Soviet citizen from the entanglements of the bourgeois capitalist self.[8] Conversely wallpaper, lace, decorative elements, the layering of cloths and the layering of objects in space were all seen as entrapping the individual, notably the domestically labouring woman. Such material surfaces needed to be cleaned frequently, decorated frequently, dusted, arranged, etc., taking up the precious time of women in the home and keeping them away from the public sphere. In socialist production, women were expected to participate alongside men in work outside the home. Smooth, industrialized, unadorned forms were designed to detach women from their earlier entanglements. The detachment imposed by whiteness has been further articulated in an entirely different manner by Diana Young in her discussion of the contemporary London property market. She argues that whiteness, because of its purported neutrality, produces real estate as a fluid commodity that is able to circulate more widely and freely and thereby gain value.[9] A parallel case concerns the redness of kula shells used in exchange networks in Melanesia. Although the shells are white in their natural state, they accrue value as they are traded from one set of hands to another. Body oils produce a red patina that indexes this fluidity and desirability.[10] In these two very different circumstances, whiteness and redness both index speed and fluidity, producing value as well as marking it.

Finally, Stephen Graham and Nigel Thrift (2007) have recently called for a critical re-engagement with the long-overlooked maintenance of building surfaces.[11] The care of surfaces has long been noted in feminist and Marxist traditions of scholarship, but these have yet to be expanded outside the domestic sphere and into the wider urban landscape. Elsewhere Thrift observes that new technologies, such as computers, videos, cameras and smart phones, have created a global surface of social engagement and consciousness focused on screens, which mediate most social and economic relations.[12] This is a landscape of pixelation that envelops the world. Screens, as primary means of contact and social engagement, are increasingly integrated into buildings, tools, objects, furniture and walls, creating a radically different sensorium. These digital elements are at once decorative, instrumental, communicative and mediative, an architectural fabric beyond the wildest musings of Semper back in the nineteenth century.

To this discussion of architectural surfaces I want to add consideration of the conditions that are currently emerging in Astana, the new capital of Kazakhstan. One of the most commonly discussed qualities of the place is its superficiality. The gleaming surfaces of office blocks, government edifices and apartment houses are all noted for their elaborate articulation. Many commentators have attributed extreme inauthenticity to these surfaces, hence the oft-used description of Astana as a city of *kartonavaia arkhitektura* or cardboard architecture (Figure 20). Yet this newest of capitals in post-Soviet Central Asia is the Brasília of the twenty-first century. It heralds the consolidation of a post-socialist Kazakhastani state and serves

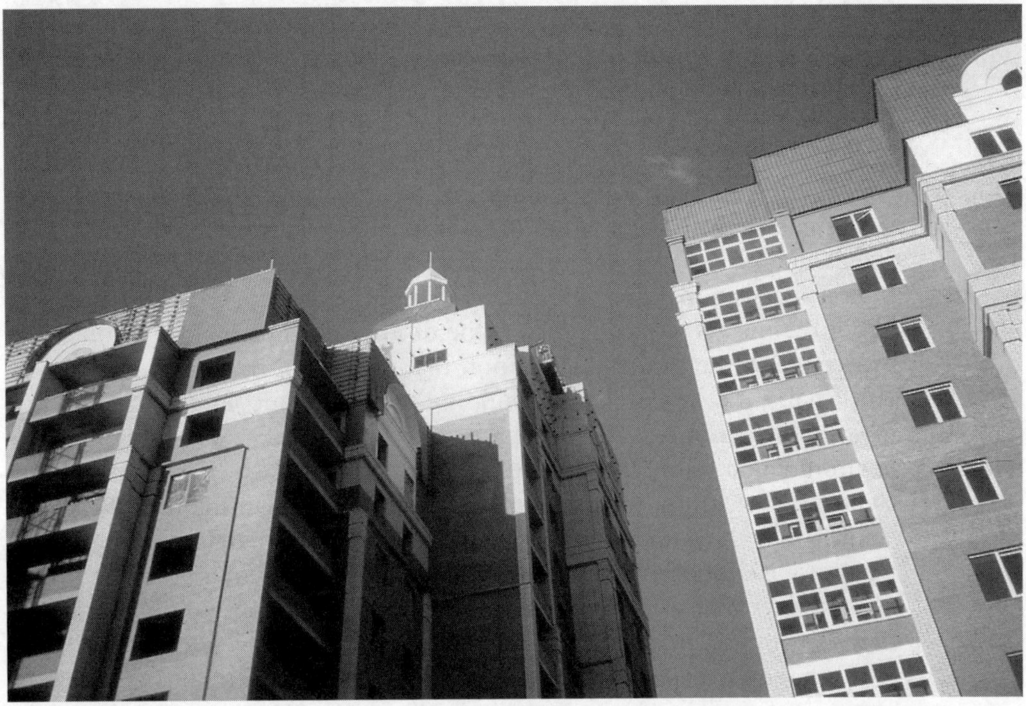

20 'Cardboard architecture', Astana

as a blueprint for nationalist development. The city was originally known as Akmolinsk, and had been a regional administrative centre since the establishment of a Cossack fortress there in 1830. In Soviet times it was known as Tselinograd, and was the capital of Northern Kazakhstan. In this guise it was the administrative staging ground for the Virgin Lands Campaign under Nikita Khrushchev, an effort to transform the arid steppes of the region into irrigated agricultural land mostly for cotton. This massive and ill-fated project diverted waterways and created one of the single greatest natural catastrophes of the twentieth century: the death of the Aral Sea.

Now Astana is the focus of another ambitious campaign, this time the remaking of the city itself. The master plan was designed by Japanese architect Kisho Kurokawa as an exemplification of his theories of cultural and urban symbiosis.[13] It has been pointedly described as the means by which the post-socialist president Nursaltan Nazarbaev is able to consolidate his power base among a young new administrative and economic elite, who are afforded political and economic advancement through the move to Astana from the former capital to the south, Almaty. With this move the president has been able to break ties with the old centres of Soviet era power in the south and their attendant clan politics and allegiances, and create a new cadre of clients in the new capital.

From imperial and Soviet times through to the present, this part of the world has always been the focus of modernizing narratives. These efforts have particularly focused on the use of building materials and their surface effects. In the imperial period one of the singular anxieties expressed was the problem caused by the use of traditional Kazakh mud brick (*samannyi*), rather than wood, as was used by the Russian peasantry. This was more than just a seemingly prosaic choice in building material. It was a question about the means by which the engine of Russian imperial expansion could proceed, and the limits placed upon that expansion by the environment – particularly the availability of building materials, and imperial ambitions to control the nomads of the steppes. Northern Kazakhstan represented the environmental borderlands between the fir forests of the north and the arid steppes of the south. Imperial colonization could only be consolidated and stabilized by the settlement of East Slavic peasants through land grants. These peasants' livelihoods were dependent on the independent farming traditions and lifeways that were facilitated by the forest, and the building materials these environmental zones provided. Without these resources Russian Slavic peasants could not sustain themselves, economically or culturally. There was also the threat of assimilation where Russian peasants began to adopt Kazakh lifeways, living in mud brick houses or yurts, adopting Kazakh subsistence practices, intermarrying and even converting to Islam.[14] Without the necessary building materials the Russian imperial project would be severely compromised. That is why in the semi-forested steppe regions of the north, Russian imperial authorities went to great lengths to provide appropriate building materials for Russian settlers. This concern was further expressed in the anxieties shown by Russian imperial medical officials who were panicked by the 'spiritual depression' of settlers and threats to their health and spirit in what were perceived to be the 'unhealthy' conditions of domestic life in mud brick structures.[15]

But this preoccupation with architectural surfaces was not just limited to anxious imperial officials. Modernizing Central Asian elites were also concerned with the quality of surfaces, which indexed the moral decline of traditional life. The dilapidated and decaying surfaces of buildings were interpreted in various ways, from the 'backwardness' of traditional Islam to the moral turpitude of pederasty.[16] A lot more was at stake than the simple maintenance of building surface, namely whole ontologies and moral economies.

With the Bolshevik revolution, this preoccupation with the social effects of surface engagements gained further strength. The backwardness of pre-revolutionary society was countered by the new modernizing forms of Soviet industrialization. This gained particular force in the context of the Virgin Lands Campaign, as the town was replanned in keeping with socialist town planning principles reminiscent of Nikolai Miliutin's archetypal socialist city, Sotsgorod. As Akmolinsk became Tselinograd, it was transformed into a provincial mid-twentieth-century city built up from mass-produced concrete panels, which displaced the old colonial districts of wooden and mud brick dwellings. As elsewhere in the Soviet Union these new forms (then gleaming) heralded an equally new set of moral and ontological orders. These industrialized building surfaces, with their easily maintained walls, indoor plumbing and clean central heating, would provide the setting for a new Soviet life (or *byt*). People living in traditional houses, whether Russian or Kazakh, were seen literally to be inhabiting a previous era of human social evolution. More privileged locals and migrants (typically European Russians and Ukrainians), by contrast, could inhabit the emergent Soviet future.

These contrasts in built form can be understood in terms of Susan Buck-Morss's opposition between socialist time and bourgeois time. Socialist time is characterized by a Hegelian notion of dialectical forward progress. The activities of the revolution are the process by which that notion of time is mastered. Bourgeois time works in the opposite direction, consolidating temporality in relation to a nationalist, typically ethnic, origin point that is fixed rather than dialectically emergent. Bourgeois time already marks a distinction from peasant culture, in which buildings serve to reproduce lifeways; yet bourgeois nationalist enterprises constructed a formalist vernacular and national style that was also static.[17] The idea of socialist Hegelian time is expressed in 'the poetics of unfinished construction', as the anthropologist Nikolai Ssorin-Chaikov puts it.[18] This poetics sees the 'decaying' and 'half built' forms that characterized Soviet built environments as indexing the future in a process of continual unfolding. The settlement, as an eternal construction site, is productive and expressive of this movement forward in time. Mathijs Pelkmans, in a similar vein, notes in reference to contemporary post-Soviet Ajaria that these unfinished forms imply a futurity that is yet to be claimed: open to everyone, and thereby establishing, paradoxically, a modicum of hope and projected stability in the unstable post-Soviet environment.[19] These surfaces, which mark an oscillation between two moments in Hegelian time, recall what Christina Kaier calls the 'dialectical objects' of earlier Constructivism, which through their striking and modernist forms indexed a movement towards socialist Hegelian time.[20] Because of their material and graphic qualities, these designs foregrounded an emergent future, as opposed to the fixed materialities of bourgeois nationalist time that consolidated a past origin. Both materialize time and space, but to diametrically opposed social effects. The half-built and decayed surfaces of socialist Hegelian time work to produce an open-ended future, whereas bourgeois nationalist institutions of heritage and museology arrange social life and surface engagements so that it remains rooted in a putatively authentic past – hence the extraordinary preoccupation with the stabilization of surfaces in architectural preservation.

In non-socialist contexts, too, unstable and decaying surfaces secure a revolutionary possibility of 'things being otherwise'. Tim Edensor has noted this in his work on the ruins of British industrialism in the wake of Thatcher era reforms.[21] Similarly the creative work of entropy is described by Caitlin Desilvey in her critique of preservationist agendas in relation to abandoned Montana farmsteads, where decaying surfaces provide a means of social engagement with the natural which challenges our notions of the cultural.[22] In all these examples, socialist, post-socialist and late-capitalist alike, surfaces and their material qualities are profoundly implicated in the terms by which morality and ontology are produced, critiqued and reformed. Both shine and decay oscillate.[23] They create foreground and background effects, which need to be understood in terms of the wider built and visual environment, and the socialities entailed through the movements those surfaces produce and enable.

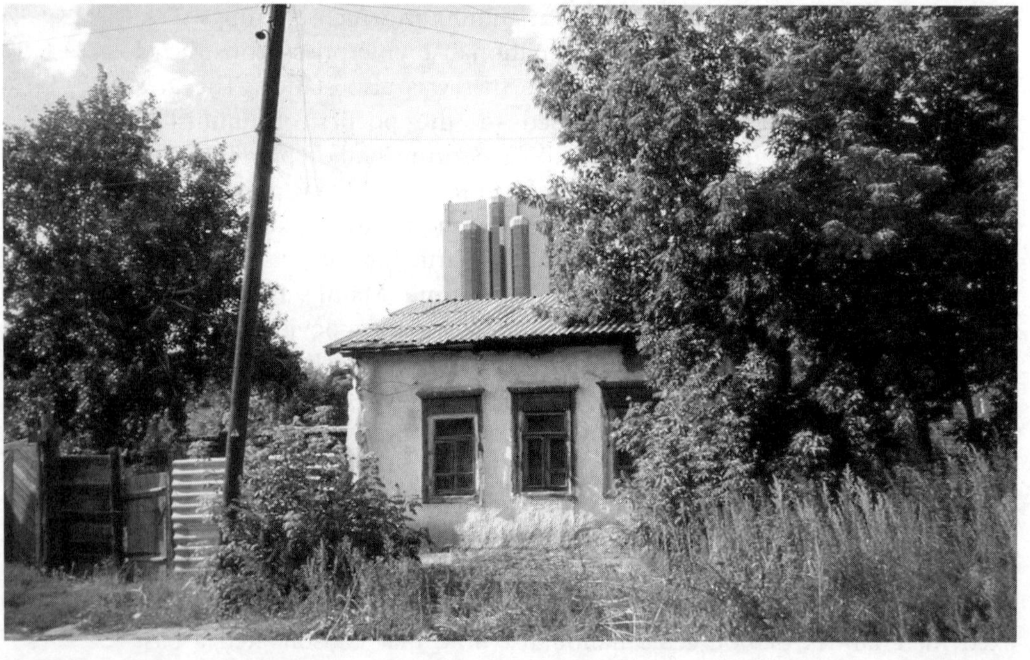

21 Old and new in contrast, Astana

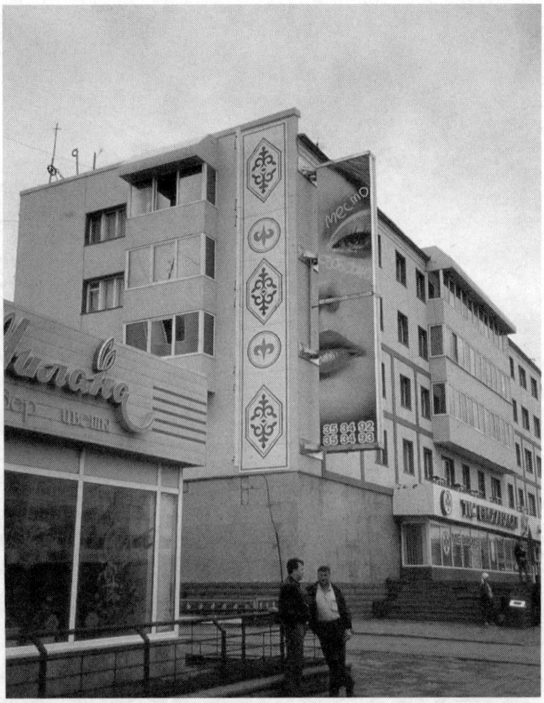

22 New cladding on Soviet-era buildings, Astana

In early twenty-first century Astana, as one might expect, this play of surfaces has an important political dimension. As the new built forms emerge and take hold they stand out strikingly against the surfaces of the Soviet and imperial eras (Figure 21). In the Hegelian manner, they strike out to inhabit a new future. But the oscillations they produce are uncertain. These façades are decried as inauthentic, tumbling down as easily as they are put up; reports demonstrate how the vinyl claddings covering up Soviet era buildings (Figure 22) are torn off during the severe winter winds. The collapsing surfaces of newly built buildings, such as the 'Prestizh' (Figure 23) – subsequently nicknamed the Titanic –began to fall apart because of cutbacks and shoddy construction, the result of deep-seated corruption plaguing the construction industry and planning and political bodies. Such physical instability, rather than indexing a deferred future, instead evinces the inauthenticity of politics and social life. In the authoritarian democracy of Nazarbaev's Kazakhstan the haggard opposition at times finds an outlet in the seemingly apolitical building press. As with Central Asian modernizing elites a century earlier, these surfaces in effect constitute a public critique and 'opposition' to corruption and bribery, better than any political tract could.

At the same time, ethnic Russians who feel displaced and threatened by the Kazakh minority that has assumed political and economic primacy

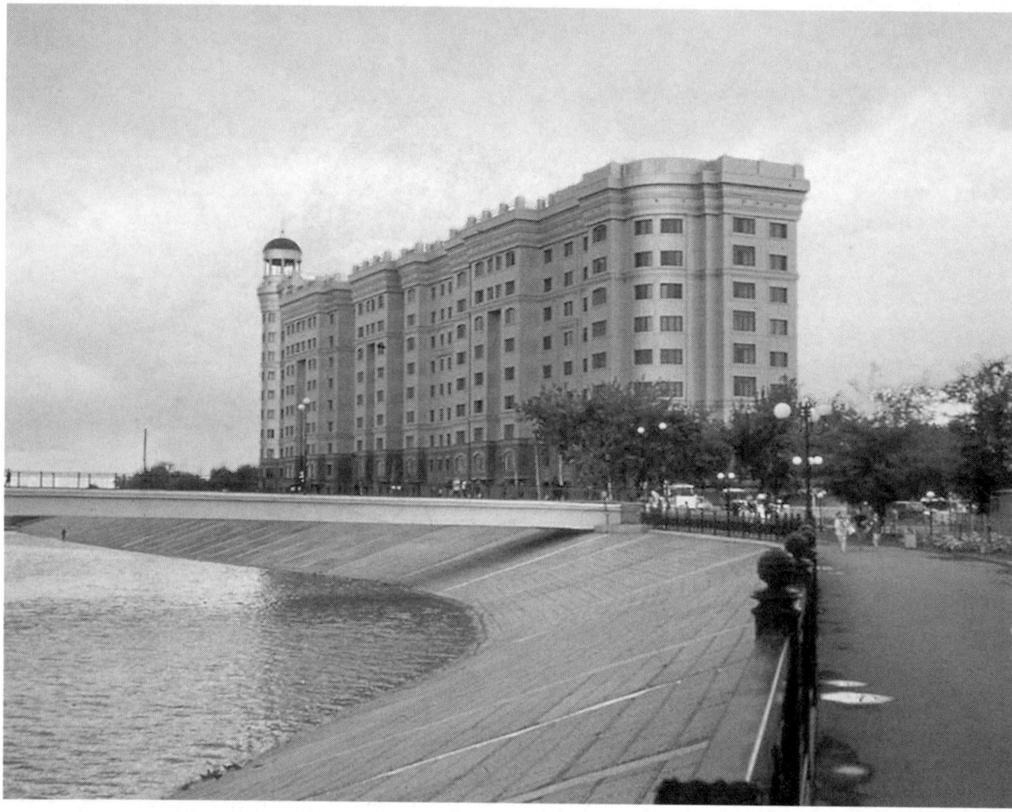

23 Prestizh, Astana

in the wake of the collapse of the Soviet Union are reassessing the wooden architectural forms of their ancestors, who first settled in the region close to two hundred years ago. In the 1980s these structures were the unvalued vestiges of an old way of life, superseded by the modern Soviet and socialist forms of what was then Tselinograd. Now they are being refigured in terms of a newly emergent preservationist discourse, which once again seeks to consolidate the origin point of Russian presence. They are no longer subsumed within the forward-looking Hegelian terms of socialist time and materiality. This architecture is recast in the pseudo-nationalist terms of territorial identification. Preservation efforts emphasize historic fabric, focusing on the physical consolidation of building surfaces. This counters the visualist narrative of the new post-socialist forms that Kazakhstani elites are producing and inhabiting. Again these two registers and their surfaces work in very different ways. Ethnic Russians attempt to ground their enduring presence through historicist architecture, while the emergent

Kazkhstani forms of the new capital project an imminent alternative future, which will come with or without ethnic Russians. In line with this ideology, preservationist efforts themselves come under attack. Much profit is to be made through development and construction, so records of older buildings disappear and houses are subject to frequent arson attacks. Meanwhile, the mud brick structures of ethnic northern Kazakhs enjoy almost no protection, certainly not from the Russians who have always distanced themselves from these forms, nor by ethnic Kazakhs for whom they serve as an index of 'backwardness'.

Recently Mateusz Laszczkowski has studied the maturing architecture of the new capital and further elaborates on the contrasting qualities of these new forms, the different kinds of national futures and pasts they constitute.[24] He notes how new media function to animate these contrasts and their social effects. In particular he notes how the pixelations of these new surfaces, which from the ground appear flat and cartoon like, exhibit all the qualities of 'camera readiness' that can be identified with the postmodern.[25] Digital print, screens, billboards and television images serve to proliferate these surfaces and marshal the population in a complicitous national building campaign that sustains the legitimacy of authoritarian governance. A new oscillation occurs here, this time between the embodied experience of walks through the new city which emerges on the left bank of the river Ishym, and the digitized images of that cityscape which circulate through various media both within the national territory and without (see Thrift on 'screen-ness').[26] Laszczkowski describes a new anxiety, one that is increasingly individualized, mediated as it is by both the perceptual experiences of walking within the urban surface-scape and the experiences of promiscuous personal media. This oscillation between the material and the virtual, Laszczkowski argues, produces a sense of individual failure, an inability to actually inhabit the promise or 'myth' of Astana. He reports that some inhabitants don't feel they actually live in the city at all. One informant observes that his room in the old socialist part of the city looks and feels nothing like the gleaming surfaces along the river Ishym, where the new city is emerging. It is this contrast, rather than the non-appearance of a promised future or imagining of 'things otherwise', that amounts to the failure of Astana as a built environment – what one might call a failure of neo-liberal personhood. The new architectural surfaces and their pixelated projections promise a normality, cast in legitimate terms of personhood, but they are complicitous with the modernizing nationalist project. They implicate, rather than an institutional and structural failure of political life, a pervasive failure of the self. Such seemingly superficial surfaces have considerable depth indeed.

Notes

1 A. Vidler, 'Diagrams of diagrams: architectural abstraction and modern representation', *Representations*, 72 (Fall 2000), 1–20.

2 V. Buchli, 'Material culture, digitization and the problem of the artefact', *Techniques et Culture*, 52 (2010), 212–231.

3 H. Heynen, *Architecture and Modernity: A Critique* (Cambridge, Massachusetts: MIT Press, 1999).

4 M. Carpo, *Architecture in the Age of Printing* (Cambridge, Massachusetts: MIT Press, 2001).

5 G. Semper, *The Four Elements of Architecture* (1851) (Cambridge: Cambridge University Press, 1989), p. 104.

6 G. McCracken, 'Homeyness', in E. Hirschman (ed.), *Interpretive Consumer Research* (Provo, Utah: Association for Consumer Research, 1989), pp. 168–183.

7 A. Gell, *Art and Agency: An Anthropological Theory* (Oxford: Oxford University Press, 1988).

8 V. Buchli, *An Archaeology of Socialism* (Oxford: Berg, 1999).

9 D. Young, 'The material value of colour: the estate agent's tale', *Home Cultures*, 1:1 (2004), 5–22.

10 N. Munn, *The Fame of Gawa: A Symbolic Study of Value Transformation in a Massim (Papua New Guinea) Society* (Cambridge: Cambridge University Press, 1986).

11 S. Graham and N. Thrift, 'Out of order: understanding repair and maintenance', *Theory, Culture and Society*, 24:3 (2007), 1–25.

12 N. Thrift, 'Beyond mediation: three new material registers and their consequences', in D. Miller (ed.), *Materiality* (Durham, North Carolina and London: Duke University Press, 2005), pp. 231–255.

13 K. Kurokawa, *Each One a Hero: The Philosophy of Symbiosis* (Tokyo: Kodansha International, 1997).

14 R. P. Geraci, 'Going abroad, or going to Russia?', in R. P. Geraci and M. Khodarkovsky (eds), *Of Religion and Empire: Missionaries, Conversion and Tolerance in Tsarist Russia* (Ithaca, New York: Cornell University Press, 2001), pp. 274–310.

15 V. Buchli, 'Astana: materiality, and the city', in V. Buchli, C. Alexander and C. Humphrey (eds), *Urban Life in Post-Soviet Asia* (London: Routledge, 2007), pp. 40–69.

16 A. Khalid, 'Representations of Russia in Central Asian Jadid discourse', in D. R. Bower and E. J. Lazzerini (eds), *Russia's Orient* (Bloomington: Indiana University Press, 1997), pp. 188–202.

17 S. Buck-Morss, *Dreamworld and Catastrophe* (Cambridge, Massachusetts: MIT Press, 2000).

18 N. Ssorin-Chaikov, *The Social Life of the State in Subarctic Siberia* (Stanford: Stanford University Press, 2003).

19 M. Pelkmans, 'The social life of empty buildings: imagining the transition in post Soviet Ajaria', *Focaal: Journal of Global and Historical Anthropology*, 41 (2003), 121–135.

20 C. Kaier, *Imagine No Possessions: The Socialist Objects of Russian Constructivism* (Cambridge, Massachusetts: MIT Press, 2005).

21 T. Edensor, *Industrial Ruins: Space, Aesthetics and Materiality* (Oxford: Berg, 2005).

22 C. Desilvey, 'Observed decay: telling stories with mutable things', *Journal of Material Culture*, 11:3 (2006), 318–338.

23 See P. Cox Miller, *The Corporeal Imagination* (Philadelphia: University of Pennsylvania Press, 2009).

24 M. Laszczkowski, 'Building the future: construction, temporality, and politics in Astana', *Focaal: Journal of Global and Historical Anthropology*, 60 (2011), 77–92.

25 G. Adamson, and J. Pavitt (eds), *Postmodernism: Style and Subversion, 1970–1990* (London: Victoria and Albert Museum, 2011).

26 Thrift, 'Beyond mediation'.

PART II

TEXTILES, SKIN

8 ✧ 'No less dangerous to use than that of a Basilisk':[1] skin and seventeenth-century English embroideries

Mary M. Brooks

Seventeenth-century English embroideries, made in girls' schoolrooms or by non-professional women at home, present miniature worlds that include elaborately worked representations of women as queens, fashionable ladies, goddesses or shepherdesses in Biblical, classical or allegorical frameworks. Made using flat stitches or elaborate raised work combining silk and metal threads with semi-precious stones, many survive as flat panels although some were made into cushions while others were mounted professionally on boxes or in frames for expensive mirrors.[2] Possibly made by women working together and apparently displayed in intimate female spaces in the house although also functioning as an indicator of female skills, these embroideries seem to have been kept away from the male gaze.[3] Scholars have explored their imagery and techniques, two rich and rewarding approaches to understanding the embroideries' cultural and physical complexities.[4] This chapter, however, focuses on one tiny aspect of these representations of women – their skin, exposed in variable degrees following the complex conventions surrounding nakedness. Skin is a meeting of the world and the body, a surface engaging with both.[5] It contains but also connects, a living, complex, permeable, changeable and vulnerable boundary which undergoes physiological changes, expresses emotions and receives impressions. Skin may be treated as a canvas and its colour and texture modified to enhance or repel. Ideas about skin are deeply embedded in language as metonyms for identity.[6] The amount of skin women expose varies according to – or in defiance of – contemporaneous fashions and social and moral codes. In these embroideries, faces, necks, cleavages, breasts, nipples and navels are all represented – and have gone largely unremarked and undiscussed. This chapter seeks to start an exploration of the possible constructions, both physical and metaphysical, of this representation of skin.

Covering and uncovering skin

It is useful first to explore the range of women depicted, whether clothed or unclothed, in order to understand the context and impact of revealed female skin. Hannah Woolley, author of household manuals and teacher of embroidery and 'lady-like' skills, was one of the few seventeenth-century women to write about embroidery. She is probably unique in writing about design and the role of the pattern drawer, the professional who drew out designs on fabrics from a range of printed sources:[7]

> I have seen such Ridiculous things done in Work, as is an abomination to any Artist to behold, As for Example, you may find in some Pieces, *Abraham* and *Sarah*, and many other Persons of Old time, Cloathed as they go now adaies … Let all Ingenious Women have regard, that when they work any Image, to represent it aright. First let it be Drawne well, and then observe the Directions which are given by Knowing Men. I do assure you, I never durst work any Scripture Story without informing myself from the Ground of it; nor any Story, or single Person, without informing myself both of the Visage and Habit;[8]

These instructions, stressing accuracy and appropriateness in depicting faces and clothing, were rarely observed by embroiderers. Distinct differences are evident in the depiction of garments between the source prints and in the embroideries. There was clearly a moment of choice for pattern drawers and makers alike where decisions could be made about the style of dress and, implicitly, the degree of acceptable nakedness.

The garments depicted may be broadly categorized as fashionable, stylized 'Biblical', or 'classical' dress. Exemplary Biblical women are normally shown fully dressed (Plate 16). Sarah, wife of the patriarch Abraham and a model of 'obedient, virtuous and beautiful'[9] womanhood, is often shown wearing a version of fashionable dress with a modest neckline or rather nondescript draped garments similar to those in the printed sources.[10] Esther, King Ahasuerus's second wife, is shown fashionably dressed, sometimes with a regal ermine-trimmed cloak and coronet. Real queens such as Henrietta Maria or Catherine of Braganza are depicted wearing elaborate fashionable dress, often with pearls and crowns. The Queen of Sheba is usually crowned and robed, an assertion of her regal status. Despite being an 'Ethiopian queen',[11] she is depicted with fair European skin and sometimes a marked décolletage following the fashion for generous displays of skin in elite dress and masquing costumes. Robert White's directions for his 1617 masque *Cupid's Banishment*, performed by the pupils at Ladies Hall, Deptford, before Queen Anne, specified that their white tissue gowns revealed 'their breasts naked (but adorned) with real jewels and pearls'.[12] Robert Aylett, a lawyer and religious poet, blamed such fashions for Bathsheba's fall from grace:

> ... a stranger at the Court
> Whose strange and new attyres, she did not know,
> Where Ladies naked brests, and shoulders show ...[13]

The display and depiction of naked breasts seem to have had defined limits of acceptability and may well have been exaggerated by artists in intimate miniatures.[14] Extreme décolletage was probably modified by in-filling low necklines with fine fabrics. Numerous embroideries show fashionable couples with only modest exposure of female skin.

Three Biblical women – Eve, Bathsheba and Susanna – are invariably depicted with naked breasts but with their lower bodies covered by a protective leaf or a fabric drape (Plate 16).[15] Bathsheba's complex and dubious story has resulted in tortuous interpretations as commentators and poets strove to present her as an acceptable role model while describing her undoubted attractions. N. H., the anonymous author of *The Ladies Dictionary*,[16] summarized Bathsheba's story with impressive succinctness:

> Bathsheba Wife of Uriah the Hittite, her naked beauties so Inchanted King David, who espyed her from a Turras, as she was washing in a Fountain, that he procured her Husband to be slain, and her to Wife ...[17]

The poet Francis Sabie makes Bathsheba complicit in her fate through her own exposure of her skin, describing her eager transformation from dressed to undressed:

> Then nimbly castes she off her Damaske frocke
> Her Satten stole most curiously made
> Her Partlet needle-wrought, her Cambrick smock ...[18]

Susanna is also depicted naked while bathing although her moral purity is never doubted.[19] *The Ladies Dictionary* describes her as:

> a chaste Hebrew Lady who was wrongly accused of Incontinency by the Elders and proved Innocent by Daniel ...[20]

A model of virtue, she exemplifies God's divine intervention on behalf of innocence.

Classical goddesses appear less frequently in these embroideries than Biblical women but share with them the generous exposure of the upper body and tactfully draped lower limbs. Stories from Ovid's *Metamorphoses* feature partially naked female figures including Diana, Daphne and Arethusa. Following the print sources, allegorical figures such as Charity are shown with exposed but nurturing, rather than erotic, breasts (Plate 17). The Senses are often depicted in fashionable dress and may expose generous décolletages. With personifications of the Four Continents, nakedness is used to make a point about civilized and primitive cultures. Europe and Asia are generally shown dressed in symbolic rather than fashionable clothing but Africa and America are usually partially clad (Plate 17).[21]

Fantastical creatures naturally had fantastical skin. Mermaids occupied an uneasy position between myth and reality[22] but what is significant here is the stress on the colour and quality of their skin. The mariners Hilles and Rayner reported the mermaid they saw on 15 June 1608 had '*skin* very white' and a tail 'speckled like a mackerel'.[23] Despite their ambiguous status ('mermaid' was a synonym for prostitute), images of bare-breasted mermaids adorned taverns, ships, furniture, ceramics and even children's primers.[24] Their iconography combined sexual temptation and unavailability with vanity and display:

> There sits a Mermaid on the Rock
> with comb and glass in hand.[25]

Regardless of this dubious reputation and their link with threatening, uncontrolled female sexuality, mermaids disporting themselves in rocky pools frequently appear on these embroideries (Plate 17). Their presence may act as a commentary on the main narrative.[26]

The male gaze

How can this embroidering of exposed skin and naked flesh by girls and women be understood? The inherent contradictions of the exemplary and the erotic in the female nude in prints and paintings have been explored in some depth but the context is different here.[27] These embroideries had long been considered unworthy of consideration, inferior decorative arts worked by anonymous female hands, so the meaning of the nude in this context has rarely been discussed. Relatively few partially clothed male figures, either Biblical, such as Adam or Isaac, or classical, such as Orpheus, Actaeon or Alpheus, appear. The female figures are explicitly exposed to the male gaze and often put at risk in consequence: Eve is made newly visible to Adam (and to God), David gazes at Bathsheba from his raised tower, the Wicked Elders conceal themselves to feast voyeuristically on Susanna's nakedness, Actaeon glimpses Diana bathing while Arethusa runs from Alpheus and Daphne flees Apollo.This threatening gaze contrasts with protective surveillance identified by Tara Hamling in depictions of Biblical narratives in early modern Protestant domestic interiors.[28] Whether shown still or in movement, the women are always carefully draped enabling their depiction to be chaste and demure but sexualized by the observing or pursuing male.[29] N.H.'s comment nicely highlights this contradictory position: David's illicit gaze is condemned but it is Bathsheba's nakedness which is the 'fatal' cause:

> We find by lamentable, if I may not say, fatal Experiences, that the world too much allowed nakedness in women. David sinned for being too prodigal and free of his looks, and some single glance sufficed to make him fall into sin. That Prince was holy, and Bathsheba on whom by accident he cast his Eyes, was innocent, but she was naked: David saw her in that posture, and there needed no more to make David loose his Holinesse and Bathsheba her Innocence.[30]

However, the gendered context of these embroideries acts to neutralize any potential eroticism in their depictions of naked skin. Made and used by women, they depict the male gaze but may rarely have been subjected to it.

The blazon of nakedness

The poetic objectification of the female body forms an important context for its representation in embroideries. Poems clearly had female as well as male readers, although their aims in reading and the lessons drawn may well have differed; the British Library's copy of *David's Troubles Remembered* is inscribed 'Mary Vaughan her book'.[31] Despite his lengthy account of Bathsheba's beauty while she 'sate all naked in a velvet chaire', Aylett explicitly links his description of the texture and colour of the skin of her 'Silken brest' to 'the first forbidden fruits of sinne'.[32] The poetic blazon of the female form, following Petrarchan ideals of fair hair, white skin, and red lips and cheeks, recalls the wrapped gilded metal threads, red dyes, feathers, pearls and semi-precious stones of the embroideries. George Ballard's description of Susanna in his poem *The History of Susanna* is typical:

> Her haires the golden threads …
> Her teeth like orient pearle, whose comly rows,
> Her lips (like Rubie) very seldome shows …
> Her paps two Swan-down worlds, that each containes
> Like Rivolets, bright Azure branched veines.
> Such kinde of commendation did but staine
> Her holy count'nance in a purer graine,
> Then blushing rose, then Vespers crimson skie,
> Then snow-bals tincted with vermilion dye.[33]

Shakespeare mockingly rejected analogies between female skin and nature ('If snow be white, why then her breasts are dun'[34]) but it was a dominant trope, evoking softness, smoothness and freshness:

> Her lovely cheeks resembling summer-fields,
> Which damaske-rose, and silver lilly yeelds:[35]

Other poets placed a sensuous stress on the breasts' sweet-smelling silky softness, equating them with fruit and flowers, a verbal counterpart of the motifs which so often surround the embroidered female form:

Her goodly bosom like a Strawberry;
Her neck like to a bunch of Columbines;
Her breast like lilies, ere their leaves be shed;
Her nipple like young-blossomed Jessamines.[36]

Poetic construction of the naked female form inevitably contrasts with the message given to young women about the importance of modesty in dress. Clothing is the consequence of Eve's sin, 'whereto were Cloathes first ordained, but to cover that nakednesse which sinne brought, and to skreene that shame which the effect of sinne first wrought?'[37] It was a woman's moral duty to cover her offending body:

> Eye those *rising mounts*, your *displayed breasts*, with what shamelesse art they wooe the shamefaste passenger ... : Contemne that fashion which ... makes it forgetfull of the Creator.[38]

These complex and contradictory attitudes to the display of the female body are neatly encapsulated by two entries in *The Ladies Dictionary*. Having described 'The sight of a fair Neck and pretty swelling breasts' as 'no less dangerous to use than that of a Basilisk', N. H. goes on to supply instructions on beautifying the neck.[39]

Skin deep

Peacham – the 'knowing man' referenced in Hannah Woolley's discussion of embroidery design – described how to paint the ideal complexion: 'lay on a weake colour, that done, trace out the eies, nose, mouth, and eare, with lake or red Lead'.[40] Female complexions in these embroideries broadly follow this format. They are typically constructed from pale creamy coloured silk fabric or threads. Fashionable rosy cheeks, worked in red threads which are now often sadly faded, once contrasted with these pale complexions. In mid-sixteenth-century England, the term 'complexion' had shifted from an almost medical concept, assessing the balance of the humours, to describing the colour, texture and state of facial skin. This was considered to have a special role in indicating health, character and even spiritual state. Underlying this was a changed view of the skin as a functioning, permeable and active membrane.[41] A 'white' and clear complexion was desirable, free of spots which might indicate pox or plague. The recipes in *Delights for Ladies* reveal the problems as much as the remedies: 'To take away spots and freckles', 'Skin kept white and cleere', 'To help a face that is red or pimpled.'[42] Involuntary spots break up the continuous 'immaculate' perfection of the skin. It is significant that the 'speckling' of mermaids' tails was always carefully embroidered, demonstrating their ambiguous nature through the contrast between 'the fair upper part' which 'however beauteous, ends in monster'.[43] Equally, moles and spots were read as gendered 'signification' of destiny and behaviour[44] while extreme redness raised questions about temperament, health and excessive eating and drinking. As yet, no examples have been found of any of these miniature embroidered beauties showing such distressing signs. However, not all have the fashionable pale complexion. Personifications such as Africa (Plate 17) and attendants on the Queen of Sheba may have 'dark' complexions. In painting, a 'swarthy' complexion was achieved by using 'sea coale, lampe blacke to deepen and shadow it'.[45] Embroiderers used black or brown thread. The presence of such 'swarthy' complexions inevitably raises questions about cultural constructions of 'black' skin. Significantly, Whitney's emblem *Aethiopem lavare,* showing two white women 'scouring' a 'blackamore', demonstrates the impossibility of whitening a 'swarthy' complexion.[46] This implicitly indicates associations between dark complexions and uncleanliness, disease, and the 'other'. Such associations lay behind the negative reactions to Queen Anne and her ladies when they took the roles of Ethiopian nymphs in Ben Johnson's 1605 *Masque of Blackness.* Their 'blacking up' was read as an 'ugly sight' which 'became them nothing so well as their red and white'.[47]

The naturalness of this favoured red and white complexion – the Petrarchan colours of love– was also important. In *Twelfth Night,* Olivia indignantly rejects the charge of wearing cosmetics, ''Tis in grain, sir; 'twill endure wind and weather' and Viola immediately concurs:

> 'Tis beauty truly blent, whose red and white
> Nature's own sweet and cunning hand laid on:[48]

However, attitudes to cosmetics were as contradictory as attitudes to nakedness. Nakedness – for the face – was a virtue: the surface needed to be presented honestly because there was 'Less temptation in nakedness than in artificial adornments.'[49] The immoral deceit of such artifice was intensified by using the same term 'complexion', signifying natural skin, for the creation of a false skin, 'a painted disguise'.[50] Such earthly vanity brought spiritual dangers: 'Miserable is the condition of that Creature, who, so her skin be sleake, cares not if her soule be rough.'[51] Despite moralizers attacking such 'painting',[52] women (not just the prostitutes condemned by diarist John Evelyn[53]) and men were determined to improve on nature. As N. H. put it in a fascinating elision of the surfaces of skin and textile, 'Art must be called in to work it [the skin] over with a Beautiful Embroidery.'[54] Sir Peter Lely's portrait of Catherine of Braganza shows her using the 'embroidery' of whitening cosmetics and rouge.[55] Faces were 'painted' with 'fucus', either red (crystalline mercuric sulphide) or white (burnt and pulverized bone)[56] and 'ceruse' (white lead and vinegar), a paint with two uses: 'with Painters a principal white Colour … and … much used by Women in painting'.[57]

Nipples, lips and cheeks were reddened with natural dyes (cochineal, madder or ochre[58]) or, again, the dangerous mercuric sulphide[59] which, as Margaret Cavendish pointed out, would 'print Age in their Face'.[60] In a further intertwining of the conceit of art imitating nature in the search for the ideal of beauty, Ballard's 'vermilion dye' and other red colouring agents might be used both to paint 'real' faces and to dye the red threads used to create the embroidered faces.

Women depict women

What is unusual in these embroideries is that this female flesh is being depicted by women. Few early modern women in England are recorded as painting or making self-portraits.[61] Few embroidered women relate to real women although the 'Temperance' hanging at Hardwick Hall has been linked to portraits of the Countess of Shrewsbury.[62] The half-figure 'portraits' associated with the Nelham workshop are now thought to be personifications of Flora rather than depictions of an actual embroiderer, possibly Elizabeth Coombe.[63] Nevertheless, these embroideries of women are representations of a generalized self, echoing their makers' awareness of their own bodies, faces and skins controlled by social pressures expressed through the physical constraints of dress and posture. Textiles were essential in achieving the required levels of modesty. In an interesting inversion of Bathsheba undressing, Susanna Perwich, renowned for her embroidery, is praised for her modesty achieved through layers of clothing:

> Nor naked was her back or breast,
> What was most chaste she loved best.
> Whisks, Handcherchiefs, she'd always wear,
> Where others shamelessly went bare ...[64]

Despite this, it was acceptable for embroiderers to depict naked female bodies in their work. In working the skin, they had the opportunity to realize the ideal of beauty which they themselves might not be able to achieve. Mobile, changeable human skin could be fixed and configured using stitches as skin itself can be stitched with a needle. Michel Serres notes that textiles, a gendered material, have some of the qualities of skin:

> an intermediary material which tradition leaves to the female, and is thus thought little of by philosophers ... veil, canvas, tissue, chiffon, fabric ... supple and fragile, linens and silks ... able to flutter like curtains, neither liquid nor solid ... pliable, tearable, stretchable ...[65]

Such mutable textiles normally cover the skin, but in these embroideries represent the skin itself. The soft silken skin evoked in poetic comparisons could be created in actual woven silk satin fabric, cut to shape and stitched to the ground fabric, often plain weave linen, to create faces (Plate 18). These appliquéd features could be flat or raised up over supports such as carved wood or bundles of fibres stitched to create contours. Diagonal tent stitch or Gobelin stitches were generally used on plain weave fabric, giving a geometric but still detailed effect. Satin stitch gave a smooth effect, not unlike the eponymous woven fabric. Great attention was given to the depiction of eyes and eyebrows:

> Brows of the glittering Eyes are Cupid's Groves of Pleasure, where he shelters himself from the violent Heat of the too flaming Opticks[66]

Pupils are often clearly worked in black and set in strongly defined sockets. One particular technique appears on several pieces: the oval stitched eye outline has a second line of red stitching inside the top curve. The eye socket itself is highlighted with a continuous stitch line. The eyebrow is worked as a continuous line arching high above the eye. X-radiography has made it possible to explore the construction of the faces and bodies below the worked skin. The radiograph of the raised work embroidery *Solomon and the Queen of Sheba*[67] shows the bundles of threads below the satin 'skin' as white areas. The threads are criss-crossed to form foreheads, grouped in circular bundles to create cheeks and tightly packed together in oblongs for noses. Hidden details of the mermaid's construction become visible. Under the smooth appliquéd silk satin 'skin', the face and body are revealed, with the breasts and belly constructed from stitched-down circles of fabric (Plate 19).

The skull beneath the skin

Like real skin, textile degrades. The physical effects of light, humidity, handling and pests distort the effect of the painstaking embroidery. Fading destroys desirable rosy cheeks, fibre degradation destroys the clarity of the fabric's surface; constructed noses break down and the black silk used to create eyes and eyebrows falls out (Plate 18). The 'skin' can become abraded and detached as the upper layer of the satin weave is lost, resulting in the loss of features and even whole faces, exposing the underlying padding as a tangle of silk threads. Such degradation is accompanied by a loss of meaning and identity. Rather than delight and wonder, the reaction becomes one of disgust and incomprehension. This echoes the disgust often aroused by actual decayed and formless human skin. The degradation of the embroidered object is not just a loss of aesthetic and physical integrity and financial value but has an emotional charge as well, especially where the loss of a face is concerned.[68] This is more disturbing than merely loss of part of the textile surface – the object's 'skin': its loss reflects our own sense of mortality.

Conclusion

The life of the skin is varied and complex, reflecting contemporary preoccupations with beauty, cleanliness and health as well as social conventions and cultural attitudes to the display of the female body. Engaging with skin offers a new approach to engaging with the makers of seventeenth-century embroideries who created numerous representations of the subtleties of female skin to set beside the painted and carved surfaces made by male artists.

Acknowledgements

Dinah Eastop, Consultant, Conservation and Heritage Studies; Michael and Elizabeth Feller; Jacqueline and Richard Holdsworth; Mark Norman, Ashmolean Museum of Art and Archaeology; Sonia O'Connor, University of Bradford and the editors for constructive feedback.

Notes

1 N. H., *The Ladies Dictionary* (London: John Dunton, 1694), p. 313.

2 Hannah Smith helpfully left a note with her box explaining the mounting of her embroidery: 'I made an end of my cabinete at Oxford … and my cabinet was mad up in the yere of 1656 at London.' (Whitworth Art Gallery, Manchester T.41.1954). The Feller Collection contains examples of a raised work mirror frame (F19) and cushion (F62); M. M. Brooks, E. Feller and J. Holdsworth, *Michael and Elizabeth Feller: The Needlework Collection I* (Guildford: Needleprint, 2011), pp. 86–87, 98–99. Other collections referenced in this paper are those of the Ashmolean Museum of Art and Archaeology, Oxford and the Whitworth Art Gallery, Manchester.

3 Lady Hoby frequently did needlework with her maids while Margaret Cavendish recalls her sisters 'working' together; see J. Moody, *The Private Life of an Elizabethan Lady: The Diary of Lady Margaret Hoby 1599–1605* (Stroud: Sutton, 2001), p. 159; S. Bowerbank and S. Mendelson (eds), *Paper Bodies: A Margaret Cavendish Reader* (Peterborough, Ontario: Broadview, 2000), p. 45. Naomi Tarrant reported a rare reference to 'an embroidered cabinet' in Lady Katherine's room in the 1681 Hamilton Palace inventory: N. Tarrant, *Textile Treasures* (Edinburgh: National Museums of Scotland Publishing, 2001), p. 39.

4 A. Morrall and M. Watt, *Twixt Art and Nature* (New York and London: Bard Graduate Centre, Metropolitan Museum and Yale University Press, 2008); X. Brooke, *Lady Lever Art Gallery: Catalogue of Embroideries* (Stroud: Sutton/National Galleries and Museums on Merseyside, 1992); Y. Hackenbroch, *English and other Needlework Tapestries and Textiles in the Irwin Untermyer Collection* (London: Thames and Hudson, 1960).

5 M. Serres, *Les Cinq Sens* (Paris: Hachette, 1998), p. 97.

6 C. Benthien, *Skin on the Cultural Border* (New York: Columbia University Press, 2002), pp. 17–24.

7 The pattern drawer in *The Fayre Mayde of the Exchange* (1607, attributed to Thomas

Heywood) drew 'bespoke' designs on a handkerchief for the love-lorn heroine Phillis Flower (Act II, Scene ii).

8 H. Woolley, *A Supplement to the Queen-like Closet* (London: Richard Lownds, 1674), pp. 192–193.

9 N. H., *Ladies Dictionary*, p. 444.

10 For example, *Abraham entertaining Angels* (Ashmolean WA 1947.191.314) is based on a print from Gerard de Jode's 1585 *Thesaurus Sacrarum Historiarum Veteris Testamenti*; see M. M. Brooks, *English Embroideries* (Oxford: Ashmolean Museum/Jonathan Horne, 2004), pp. 32–33.

11 A. Lanyer, *Salve Deus Rex Judaeorum* (London: Richard Bonian, 1611), line 1569.

12 S. P. Cerasano and M. Wynne-Davies, *Renaissance Drama by Women* (London: Routledge, 1996), p. 88.

13 R. Aylett, *David's Troubles Remembered* (London: Richard Hodgkinsonne, 1638), p. 17.

14 For artists enhancing décolletage, see A. Ribeiro, *Fashion and Fiction: Dress in Art and Literature in Stuart England* (London: Yale University Press, 2005), pp. 130, 136–138; E. Chirelstein, 'Lady Elizabeth Pope: the heraldic body', in L. Gent and N. Llewellyn (eds), *Renaissance* Bodies (London: Reaktion Books, 1990), p. 57.

15 For an examination of the female breast depicted in painting and distinctions between the religious, anatomical and pornographic breast, see M. R. Miles, *A Complex Delight: The Secularization of the Breast 1350–1750* (Berkeley: University of California Press, 2008).

16 Considine and Brown argue these initials may conceal the authorship of the publisher John Dunton being a reversal of the initials 'H. N.' derived from an anagram of his name ('Hid unto Non[e]'); J. Considine and S. Brown (eds), *The Ladies Dictionary* (Farnham: Ashgate, 2010), pp. xiv–xv.

17 N. H., *Ladies Dictionary*, p. 41.

18 F. Sabie, *Adams Complaint: The Old Worldes Tragedie David and Bathsheba* (London: Richard Jones, 1596), unpaginated. Sabie sometimes misnames Bathsheba as 'Beersheba', a place name, suggesting he was neither a clergyman nor the schoolmaster with whom he has so long been identified: A. Stanford, 'Francis Sabie: a biographical sketch', *Huntington Library Quarterly*, 25:4 (1962), 278.

19 For a discussion of the ambiguous eroticism of Renaissance and Baroque prints of Susanna, see H. D. Russell, *Eva/Ave: Women in Renaissance and Baroque Prints* (Washington: National Gallery/New York: Feminist Press, 1990), p. 32.

20 N. H., *Ladies Dictionary*, p. 407.

21 A. Wells-Cole, *Art and Decoration in Elizabethan and Jacobean England* (London: Yale University Press, 1997), pp. 113–114, 193; A. R. Jones and P. Stallybrass, *Renaissance Clothing and the Materials of Memory* (Cambridge: Cambridge University Press, 2000), p. 151.

22 A. S. Goodman, 'The extraordinary being: death and the mermaid in baroque literature', *Journal of Popular Culture*, 17:3 (1983), 32–48.

23 Emphasis in the original. G. M. Asher (ed.), *Henry Hudson* (London: Hakluyt Society, 1860), 27, p. 28.

24 For example, the mermaid on the chair in Gower's 'Armada' portrait; A. Belsey and C. Belsey, 'Icons of divinity: portraits of Elizabeth I', in Gent and Llewellyn, *Renaissance*

Bodies, p. 14; J. Overton, *A New Book of all Sorts of Beasts: or a Pleasant Way to Teach Young Children to Reade* (1617) cited in L. Arthur, *Embroidery 1600–1700 at the Burrell Collection* (London: John Murray/Glasgow Museums, 1995), p. 78.

25 *The Praise of Saylors*, Pepys Ballads 4.197; Pepys Library, Magdalene College, Cambridge.

26 For example, a sampler depicting Susanna and mermaids (F23); Brooks *et al.*, *Feller: The Needlework Collection I*, pp. 52–53.

27 M. R. Miles, *Carnal Knowing: Female Nakedness and Religious Meaning* (Boston: Beacon Press, 1989); Russell, *Eva/Ave*; M. Warner, *Monuments and Maidens* (London: Weidenfeld and Nicolson, 1985).

28 T. Hamling, *Decorating the 'Godly' Household: Religious Art in Post-reformation Britain* (New Haven: Yale University Press, 2010).

29 Depictions of Joseph fleeing from seduction by Potiphar's wife are a rare variant.

30 N. H., *Ladies Dictionary*, p. 308.

31 Aylett, *David's Troubles*.

32 *Ibid.*, p. 19.

33 G. Ballard, *The History of Susanna* (London: William Hope, 1638), unpaginated. 'Graine' is the red dye kermes derived from the insect *Coccus ilicis*.

34 W. Shakespeare, *Shake-speares sonnets* (London: T. T., 1609), Sonnet 130.

35 Ballard, *History of Susanna*, unpaginated.

36 E. Spenser, *Amoretti*, Sonnet 64 (1595) cited in M. Yalom, *A History of the Breast* (London: Pandora, 1998), p. 78.

37 R. Brathwaite, *The English Gentlewoman* (London: Michaell Sparke, 1631), p. 7.

38 *Ibid.*, p. 8.

39 N. H., *Ladies Dictionary*, pp. 313, 317.

40 H. Peacham, *The Gentlemans Exercise* … (London: J. Browne, 1612), pp. 95–96.

41 S. Connor, *The Book of Skin* (London: Reaktion Books, 2004), pp. 19–20.

42 H. Plat. *Delightes for Ladies* (London: H. L[ownes], 1608), unpaginated.

43 R. Allestree, *Forty Sermons* … (Oxford and London: R. Scott, G. Wells, T. Sawbridge, R. Bentley and A. Swall, 1684), p. 25.

44 S. Strangehopes, *A Book of Knowledge* (London: J. Deacon, 1669), pp. 41–43.

45 Peacham, *Gentlemans Exercise*, pp. 95–96.

46 H. Green (ed.), *Whitney's 'Choice of Emblemes'* (1866) (Hildesheim, Germany and New York: Olms, 1971), p. 57.

47 Dudley Carleton to Ralph Winwood cited in B. Andrea, 'Black skin: the Queen's masques', *English Literary Renaissance*, 29:3 (1999), 265.

48 *Twelfth Night*, Act 1, Scene V, 256–258.

49 N. H., *Ladies Dictionary*, p. 316.

50 Brathwaite, *English Gentlewoman*, 'Introduction' (unpaginated).

51 *Ibid.*, p. 13.

52 For moralists inveighing against cosmetics, see P. Phillippy, *Painting Women: Cosmetics, Canvases and Early Modern Culture* (Baltimore: Johns Hopkins University Press, 2006) p. 203, n. 29.

53 W. Bray (ed.), *The Diary of John Evelyn* (New York and London: N. W. Dunne, 1901), p. 285.

54 N. H., *Ladies Dictionary*, p. 113.

55 Ribeiro, *Fashion and Fiction*, p. 240.

56 Plat, *Delightes for Ladies*.

57 N. H., *Ladies Dictionary*, p. 108; see Phillippy, *Painting Women*, p. 31.

58 Cochineal is derived from the insect *Dactylopius coccus*, madder from the plant *Rubia tinctoria* while ochre is the earth pigment iron (III) oxide.

59 Ribeiro, *Fashion and Fiction*, pp. 171–172.

60 M. Cavendish, *The Worlds Olio* (London: Martin and Allestrye, 1655), pp. 112–113.

61 Exceptions, both professional and recreational, include Levina Teerlinc (1510/20–76), see R. Strong, *Artists of the Tudor Court* (London: Victoria and Albert Museum, 1983), pp. 52–57; Arbella Stuart (1575–1615) whose self-portrait was described as a 'very well Enameld Picture', see S. J. Steen (ed.), *Letters of Lady Arbella Stuart* (New York: Oxford University Press, 1994), p. 52; Mary Beale (1632–99), see T. Barber, *Mary Beale* (London: Geffrye Museum Trust, 1999); Jane Myddelton (c. 1646–92), whose own Lely portrait (1666) conforms to the ideal of rose-cheeked beauty and exposed skin, and who was recognized as an 'accomplished amateur artist', see C. MacLeod and J. M. Alexander (eds), *Painted Ladies* (London and New Haven: National Portrait Gallery/Yale Centre for British Art, 2001), p.103.

62 S. M. Levey, *Embroideries at Hardwick Hall* (London: National Trust, 2007), p. 93.

63 Hackenbroch, *English and other Needlework*, plate 62 and p. 34; Morrall and Watt, *Twixt Art and Nature*, pp. 30–31; M. Swaine, 'John Nelham's needlework panel' and J. Nevinson, 'John Nelham embroiderer', *Bulletin of the Needle and Bobbin Club*, 65:1–2 (1982), 3–16, 17–19.

64 J. Batchiler, *The Virgins Pattern* (London: Simon Dover, 1661).

65 M. Serres, *Atlas* (Paris: Editions Julliard, 1994), p. 45 cited in S. Connor, 'Topologies: Michel Serres and the shapes of thought', *Anglistik*, 15 (2004), 105–107.

66 N. H., *Ladies Dictionary*, p. 64.

67 M. M. Brooks and S. O'Connor, '"In needlework there doth great knowledge rest": the contribution of X-radiography to the understanding of seventeenth-century English embroideries from the Ashmolean Museum of Art and Archaeology, Oxford', in S. O'Connor and M. M. Brooks, *X-Radiography of Textiles, Dress and Related Objects* (Oxford: Elsevier, 2007), pp. 237–248.

68 The attention paid to 'restoring' faces in the Hardwick hangings is a case in point; Levey, *Embroideries at Hardwick Hall*, pp. 62, 84.

9 ✧ Colour, language and light in mid-nineteenth-century women's fashion

Charlotte Nicklas

THE perception of colour was one part of a broader nineteenth-century preoccupation with vision that emerged as physiological understanding of the eye and technologies of seeing developed over the course of the century.[1] Incorporating this contemporary interest in perception, comments about colour in dress textiles filled mid-nineteenth-century writing about middle-class female fashion in Britain and the United States. This chapter examines three aspects of this subject: textual descriptions of textile colours and surfaces, references to the sciences of colour and vision, and discussions of clothing in different kinds of light. The awareness of science and vision provides an important, unexplored context in which to understand nineteenth-century dress and fashion. Change in nineteenth-century fashion in the United States and Britain has often been narrated as a sequence of different silhouettes. Cut, or form, was indeed a fundamental element of fashionability during this period, but this chapter will focus on the contemporary significance of fashionable textile surfaces.[2]

Increasing awareness of the fallibility of the human eye was a source of both fascination and unease during this period.[3] The unreliability of vision could dovetail with inconsistencies between private character and public appearance, a concern that underlay much nineteenth-century commentary about women's fashion and dress.[4] In the *Englishwoman's Domestic Magazine* in September 1873, Matilda Browne articulated how interior and exterior states should, ideally, correspond, when she exhorted her readers: 'let us show the order and purity of our minds in our outward clothing'.[5] Reality, however, did not always match this ideal of propriety, and criticism of this supposed correlation also appeared in women's magazines.[6] Fashion writing highlighted multiple ways in which the eye could be bewitched.

Most of the textual examples in this chapter are drawn from those middle-class women's magazines of the period which had the largest circulations; the *Englishwoman's Domestic Magazine* (1852–79) in Britain and *Godey's Lady's Book and Magazine* (1830–98) in the United States.[7] For the years before the introduction of the *Englishwoman's Domestic Magazine*, material from the British magazine the *Ladies' Cabinet* (1832–70) will be considered, although this publication was aimed at upper-middle-class women.[8] Although British and American women's magazines in the mid-nineteenth century displayed significant differences, especially in their treatment of issues of national identity and governance, this chapter will focus on similarities between discussions in these periodicals. They shared a language of fashion that enables consideration of the daily lived experience of dress, colour and light on both sides of the Atlantic Ocean.

Describing colours and surfaces

Although fashion plates became more common in women's magazines during the mid-nineteenth century, much of the information about the colour and texture of dress objects had to be conveyed in black text.[9] Descriptions sometimes accompanied colour or black and white images, but also appeared independently of illustrations. Words, therefore, encouraged the exercise of readers' imaginations. As Kate Flint notes, observation and imagination – 'outward and inward seeing' – were connected, although not necessarily considered equal.[10]

Writers for these periodicals employed a wide range of descriptive techniques to communicate different kinds of colour effects to their readers, using an established language of fashionable colour. Nonetheless, the difficulty of describing colours was admitted by nineteenth-century fashion writers.[11] In September 1872 Matilda Browne, the fashion writer for the *Englishwoman's Domestic Magazine*, describing some Irish poplins that she had seen, asserted: 'The softness of the colours and the brilliancy of the fabric cannot be rendered by mere description.'[12] Describing the season's brocades, a fashion writer for the *Lady's Book* in April 1863 noted, 'Mauve grounds with deep rich purple and black standing in relief, deep blue with a tiny frostwork of silver powdering the surface, and many other combinations that mock the dulness [*sic*] of the pen'.[13] Admitting the inadequacy of language in representing colour, these writers invited their readers to see or imagine these textiles for themselves. These descriptions drew readers into a world of 'luxurious fantasy' created by fashion magazines.[14]

Describing the surface sheen of silk dress textiles was particularly challenging for fashion writers, but this shiny surface constituted an important aspect of these textiles' desirability. Mary Schoeser notes that the shiny quality of silk, as well as its ability to hold bright colour, has contributed to its value across time and cultures.[15] A short piece in the *Lady's Book* in May 1861 summarized many of the appealing aspects of this 'agreeable and healthy material'. The writer noted that silk clothing 'retains

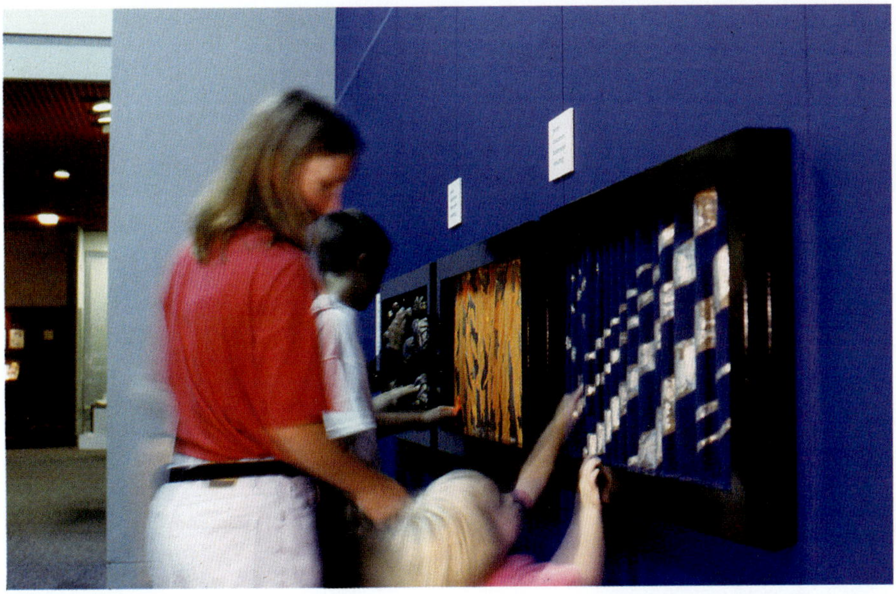

1 Frances and Ron Geesin, *Tri-Aura*

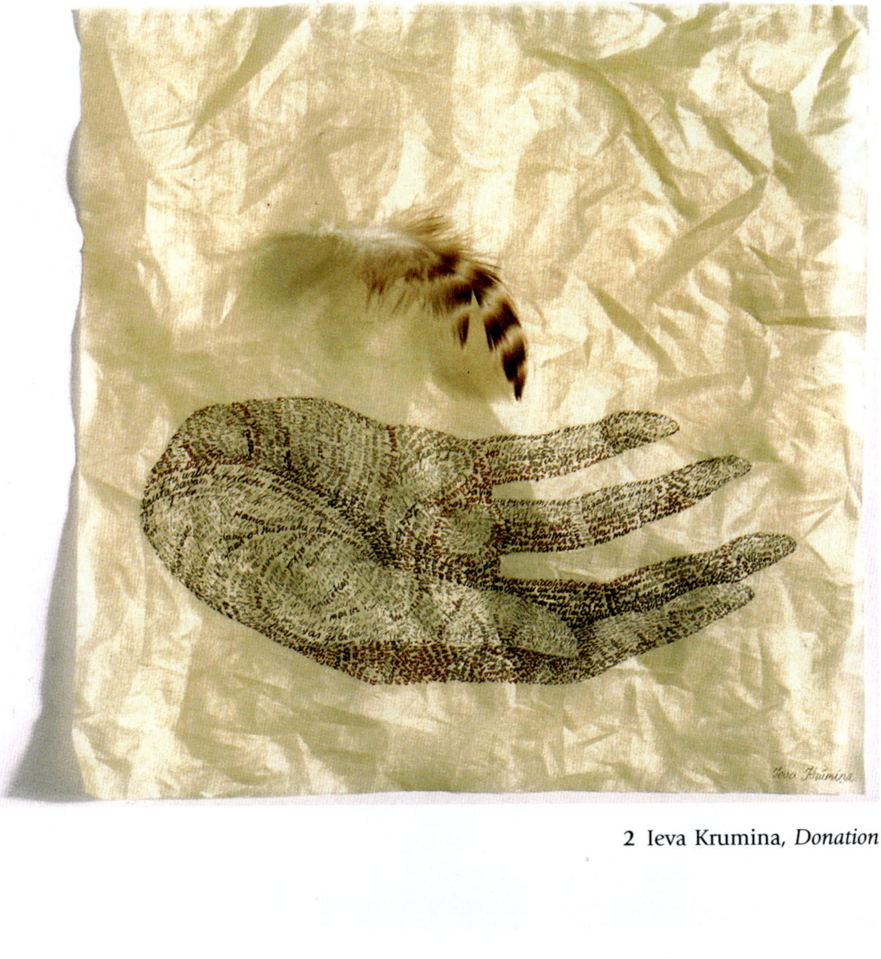

2 Ieva Krumina, *Donation*

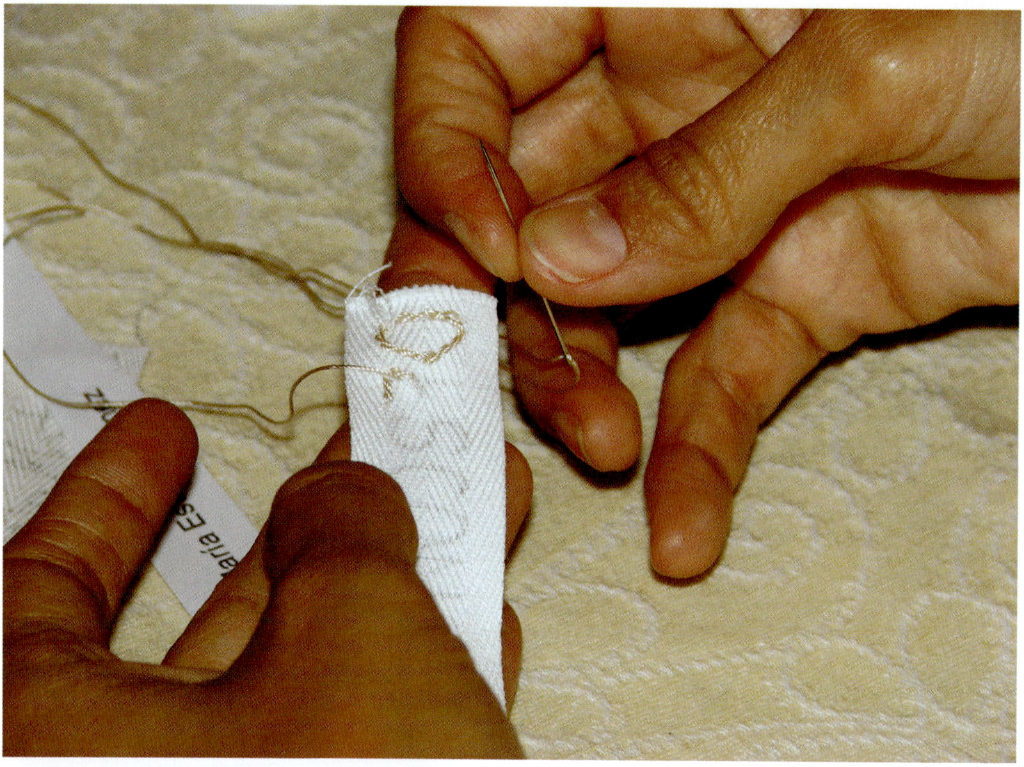

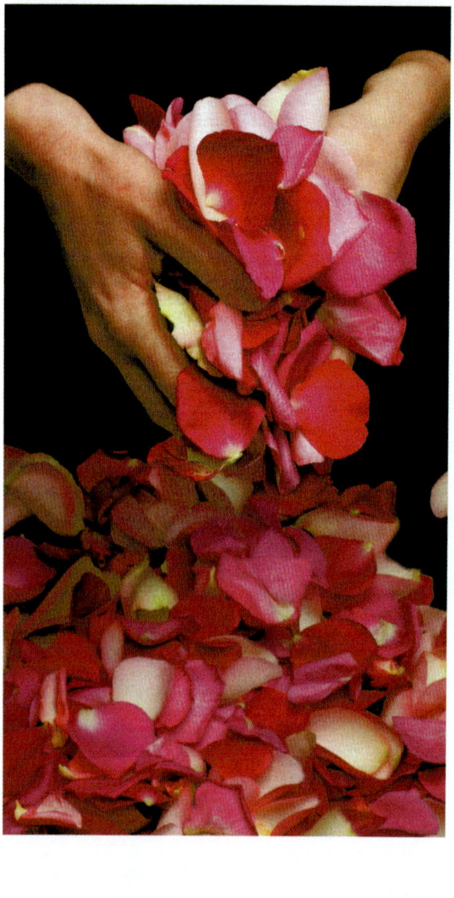

3 Lise Bjorne Linnert, *Desconocida/Unknown*

4 Liz Nilsson, *Letting Go*

5 Masayo Ave, *i-Haptic touch*

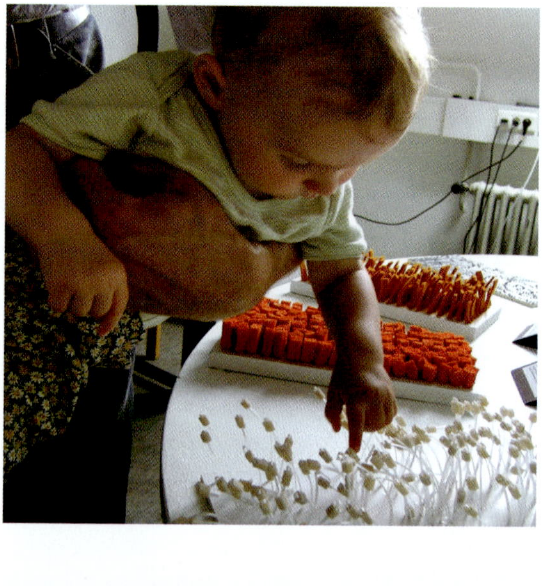

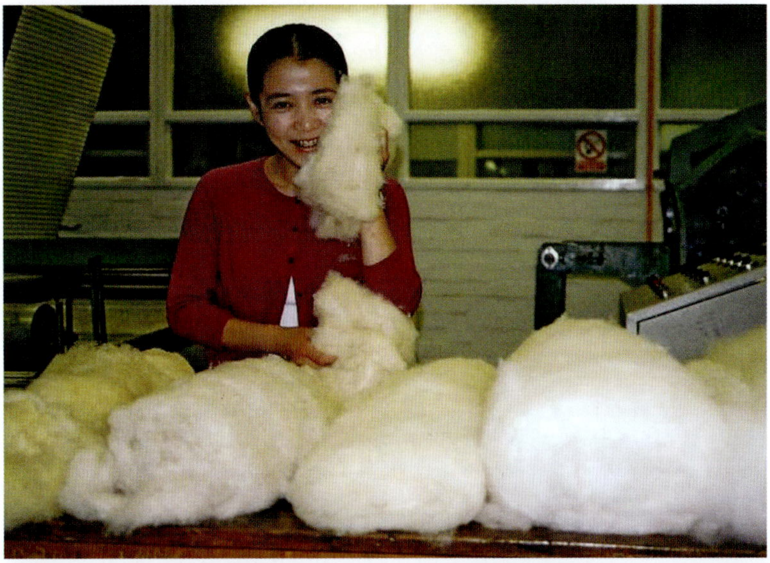

6 Naoko Yoshimoto

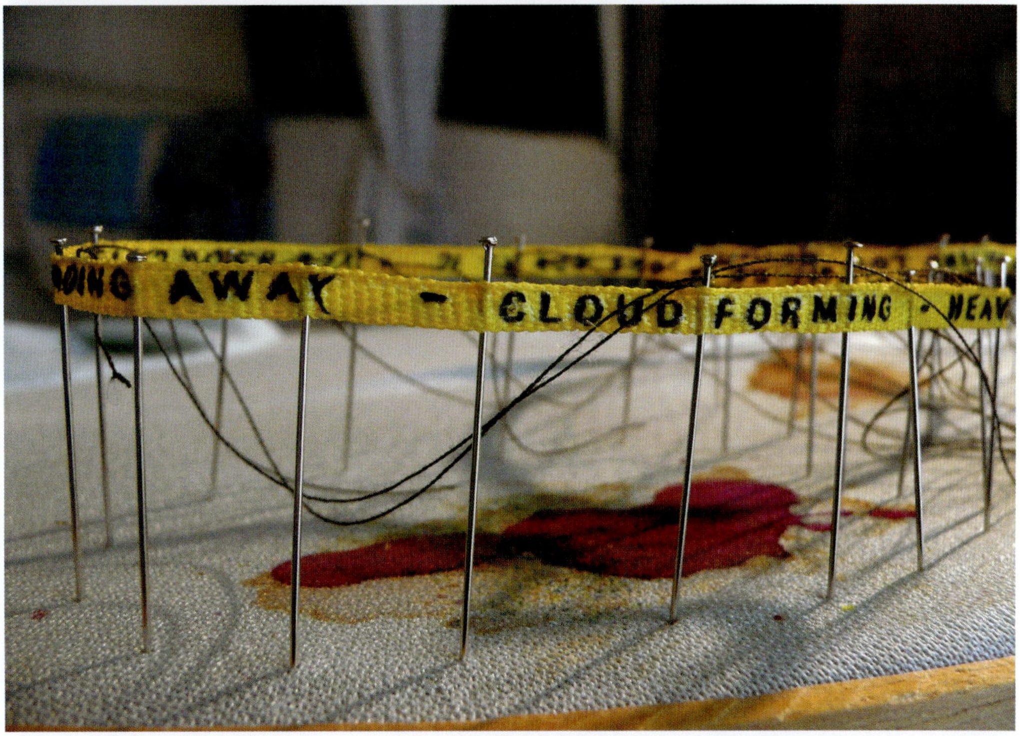

7 Shelley Goldsmith, *Outbursts 3 & 4, Revealing Evidence*

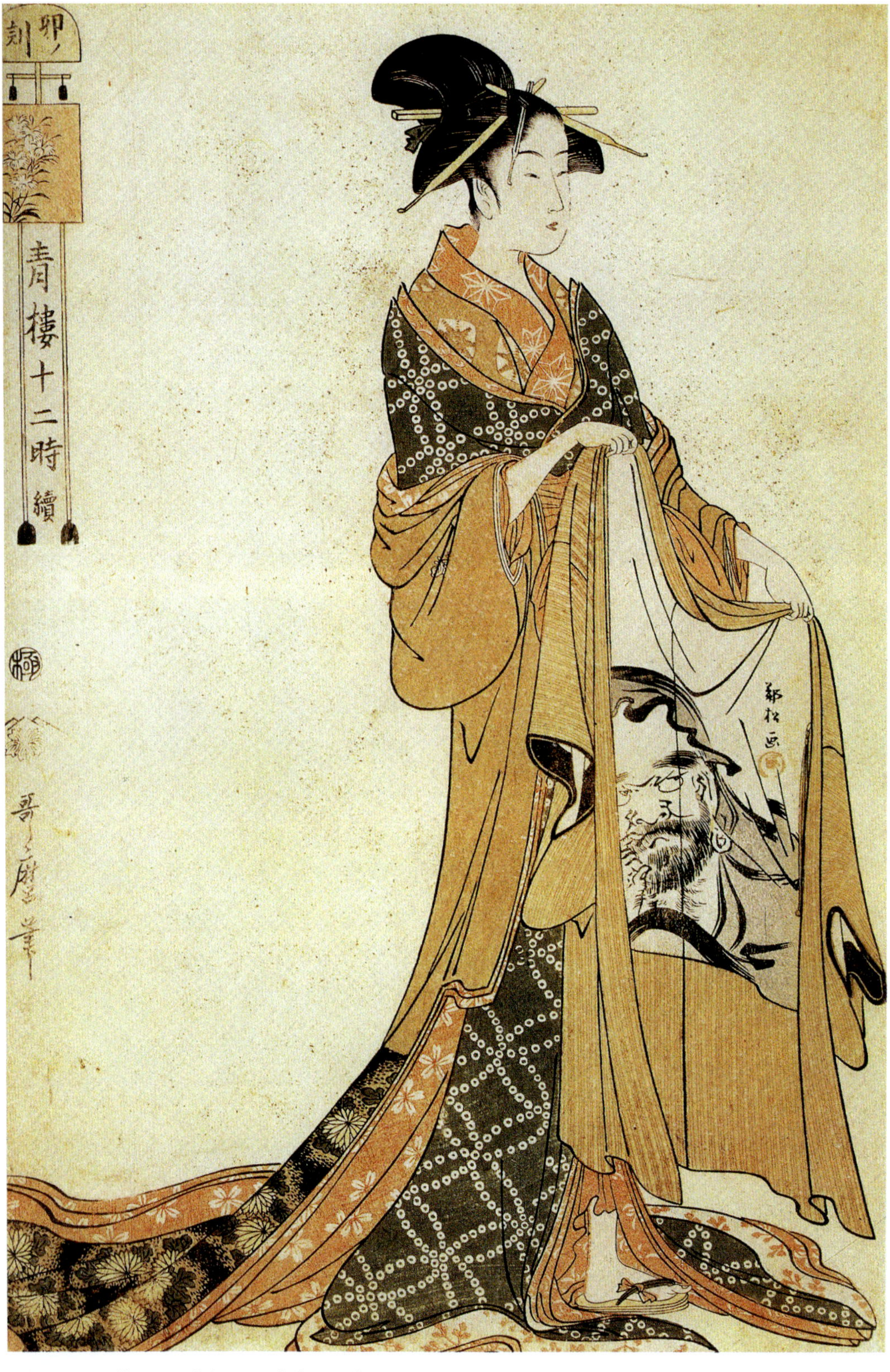

8 Utamaro, '"Hour of the Hare', from *The Twelve Hours in the Yoshiwara*, colour woodblock print, c. 1795

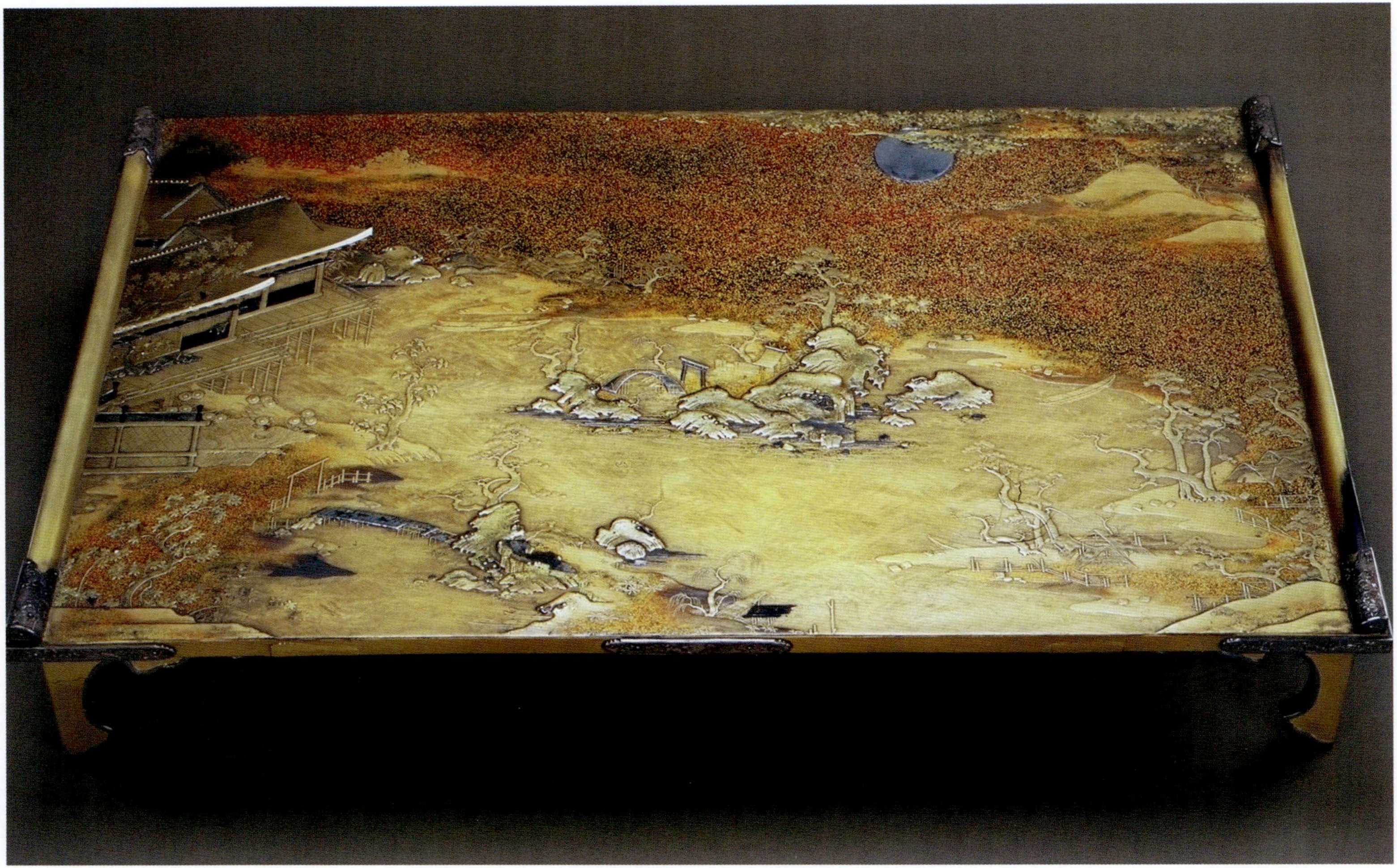

9 Makie lacquer writing table, wood, covered with gold and silver, c. 1620

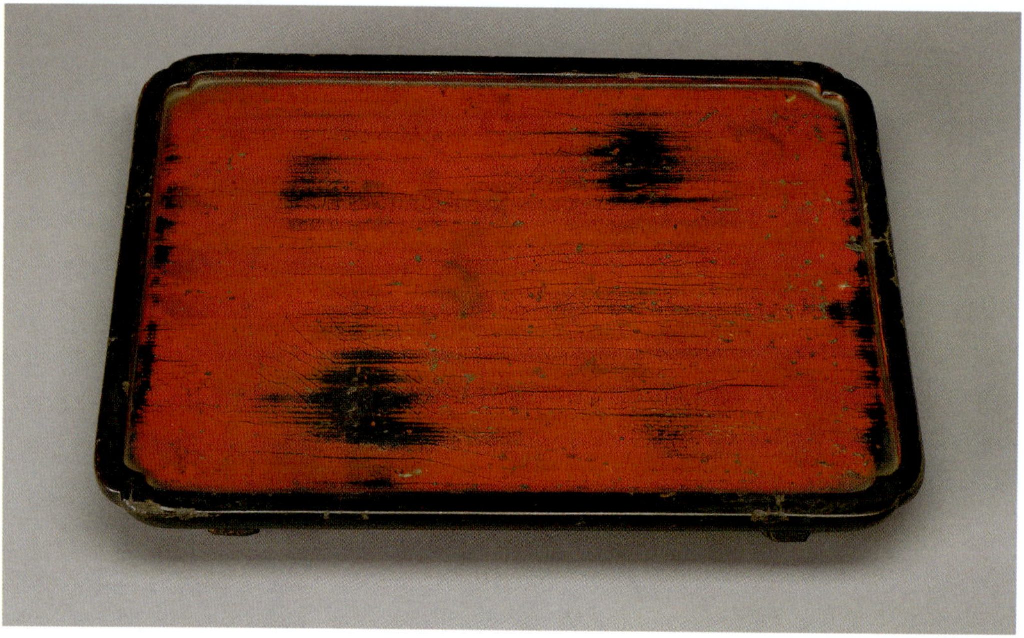

10 Negoro ware tray, lacquered wood, fourteenth century

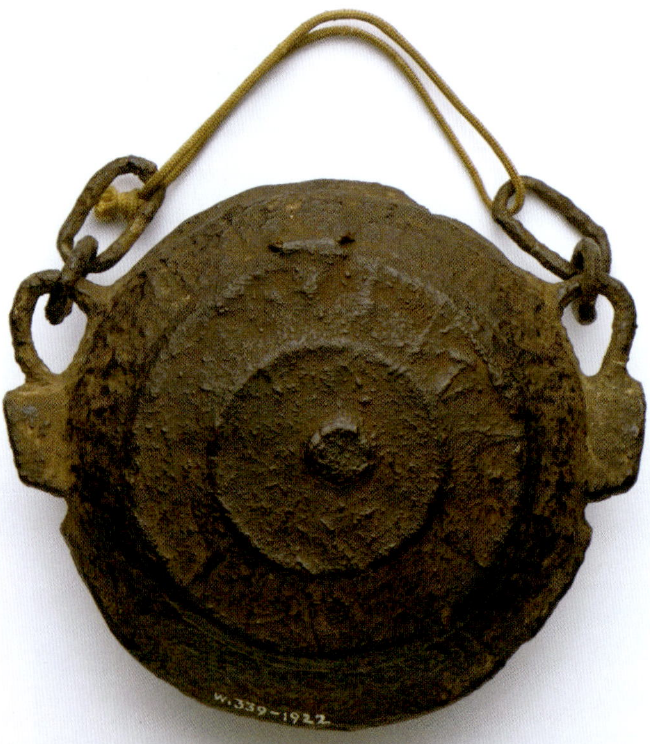

11 Ritsuô (Ogawa Haritsu), *inrô* in the form of a temple gong, eighteenth century, lacquer imitating patinated metal

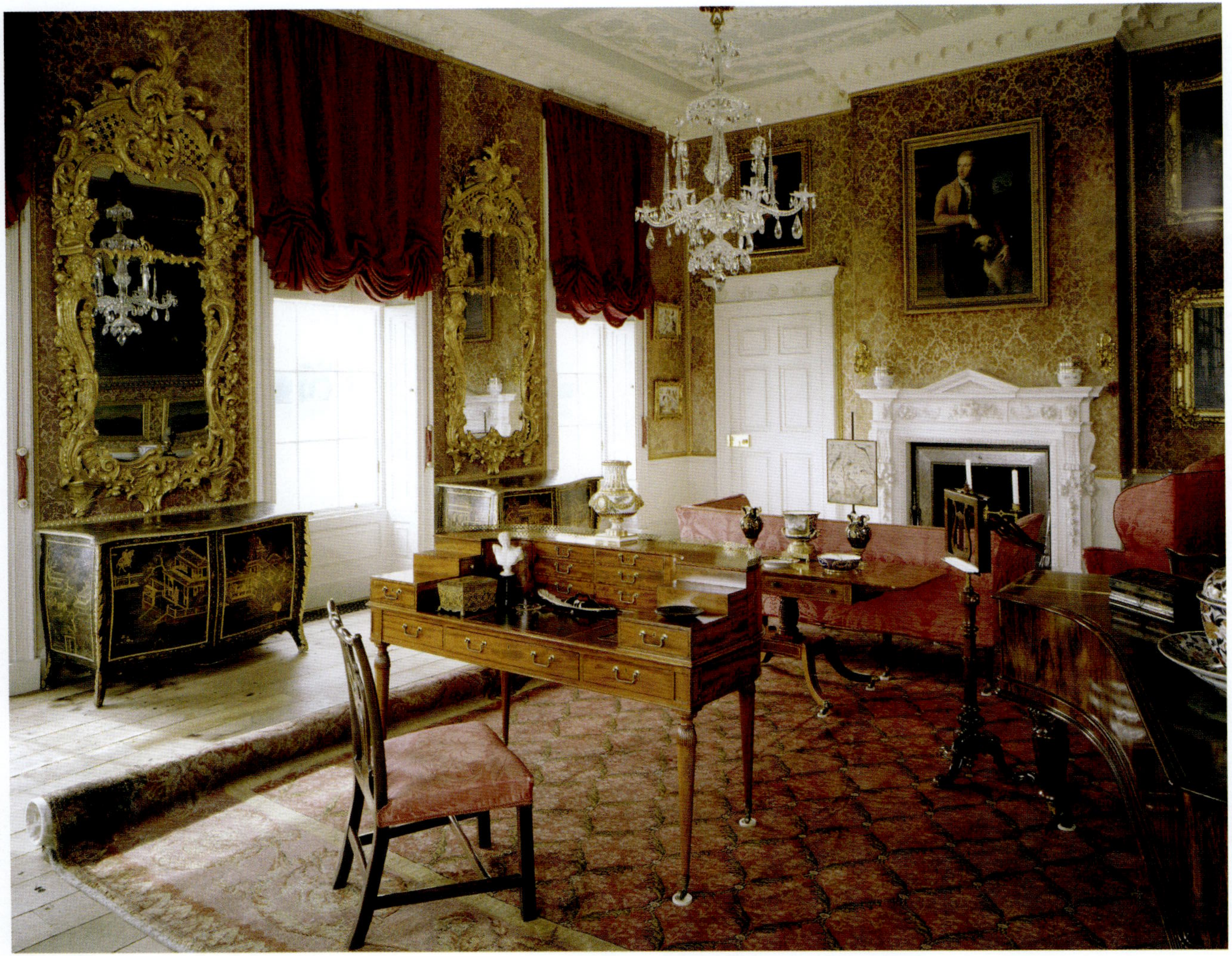

12 Red Drawing room at Uppark. Furnishings include curtains conserved by Lady Meade-Fetherstonhaugh in the 1930s along with an Axminster carpet c. 1800 and nineteenth-century faded flock wallpaper that originally matched the deep red curtains

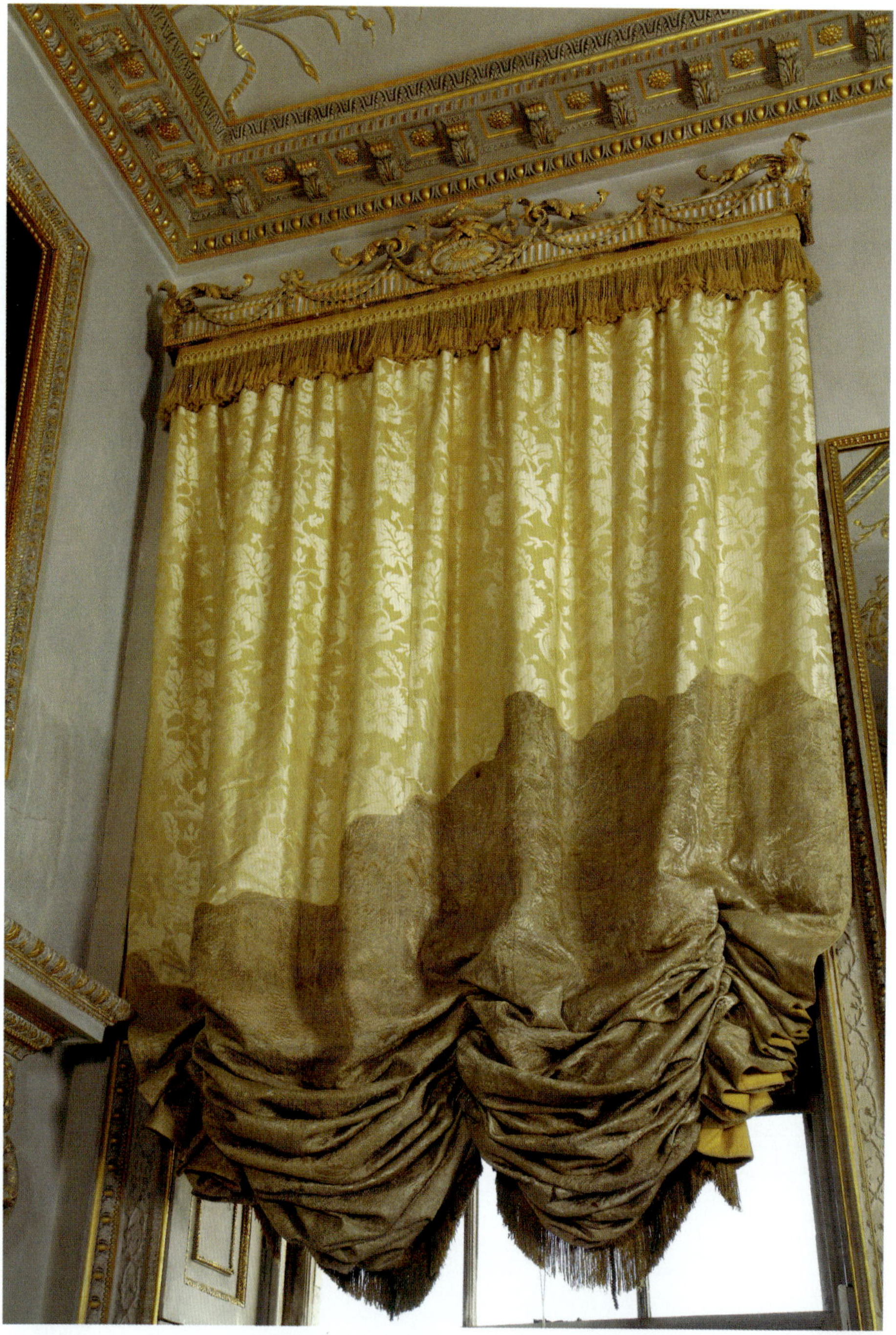

13 Curtain in Saloon at Uppark. The new textiles used to replace fire-damaged sections of window curtains are clearly visible and illustrate the National Trust's philosophy for dealing with the conservation of Uppark, to retain as much original material as possible

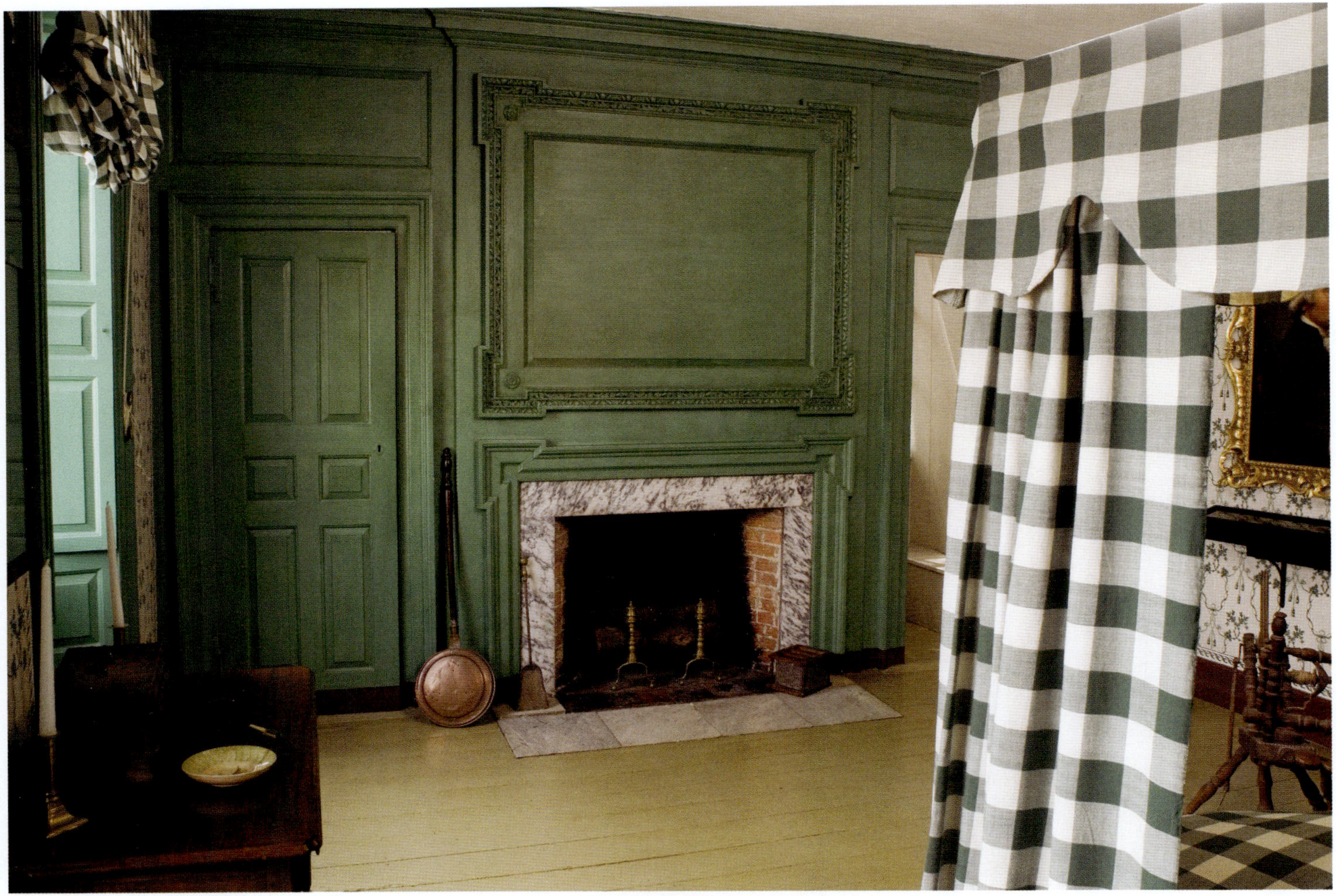

14 William Whipple's bedroom at the Moffatt-Ladd House was restored to its original appearance based on paint analysis and eighteenth-century estate inventories. The fabric for the bed hangings was specially woven at Old Sturbridge Village for use in this room

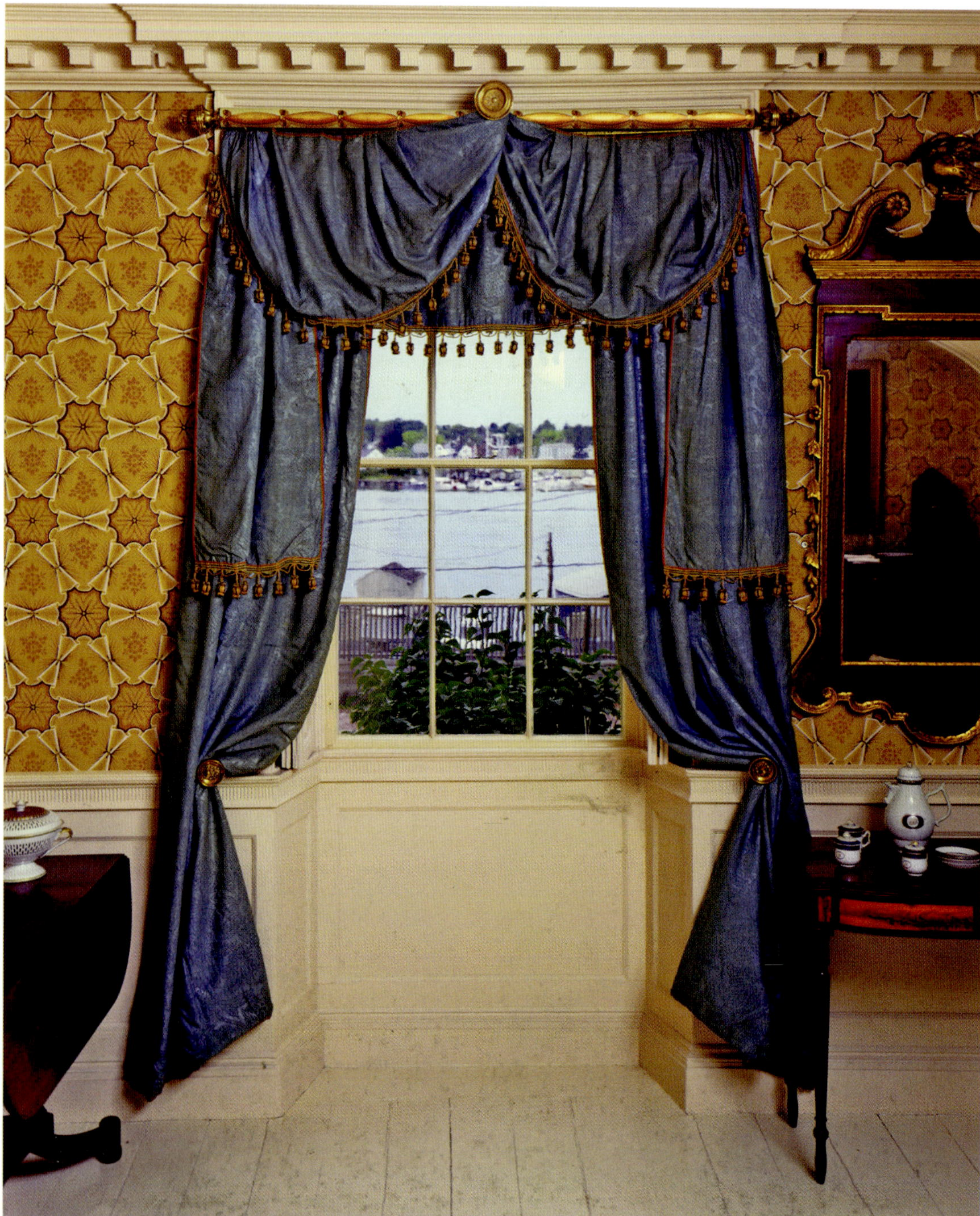

15 These early nineteenth-century curtains continued to be displayed at the Moffatt-Ladd House in its early years as a museum but were then retired and replica textiles hung in their place in the 1990s

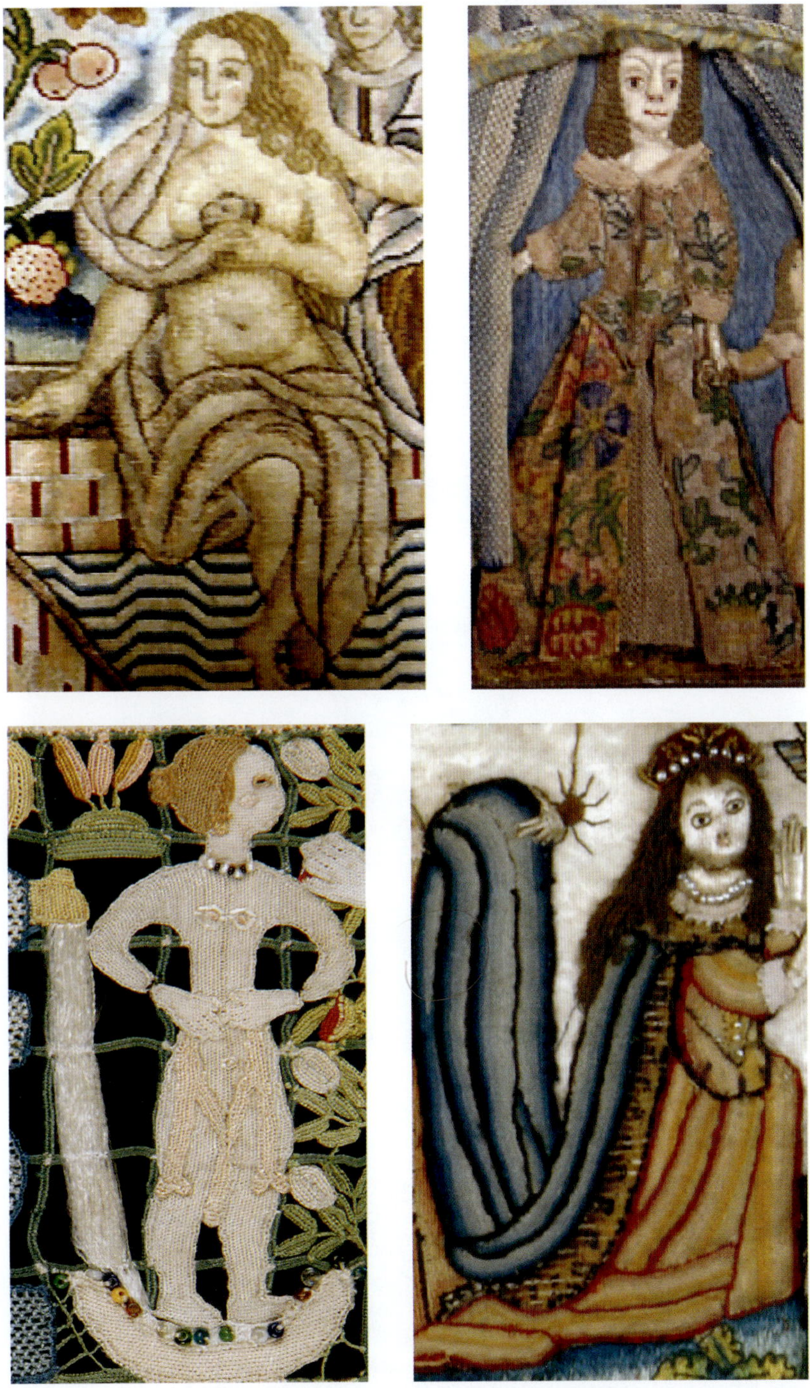

16 Dressed and undressed women (clockwise from top left): Bathsheba (detail, panel, Ashmolean Museum); Sarah (detail, box, Ashmolean Museum; note one section of the skirt is a reconstruction); Esther (detail, panel, Feller Collection); Susanna (detail, sampler, Feller Collection)

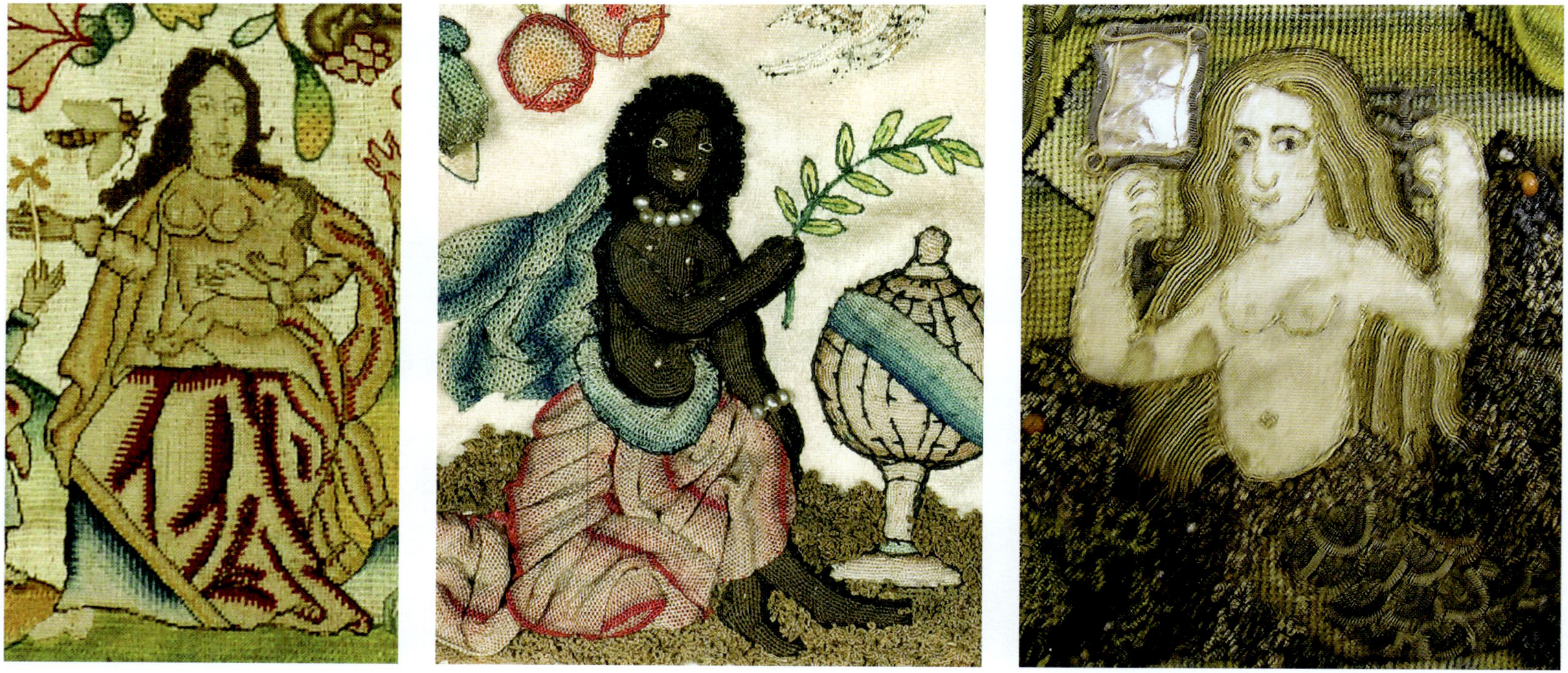

17 Allegorical, personified and imaginary women (left to right): Charity (detail, panel, Feller Collection); Africa (detail, panel, *Four Continents and Scene of Rural Life*, Feller Collection); Mermaid (detail, *Proclamation of the Young Solomon*, Ashmolean Museum)

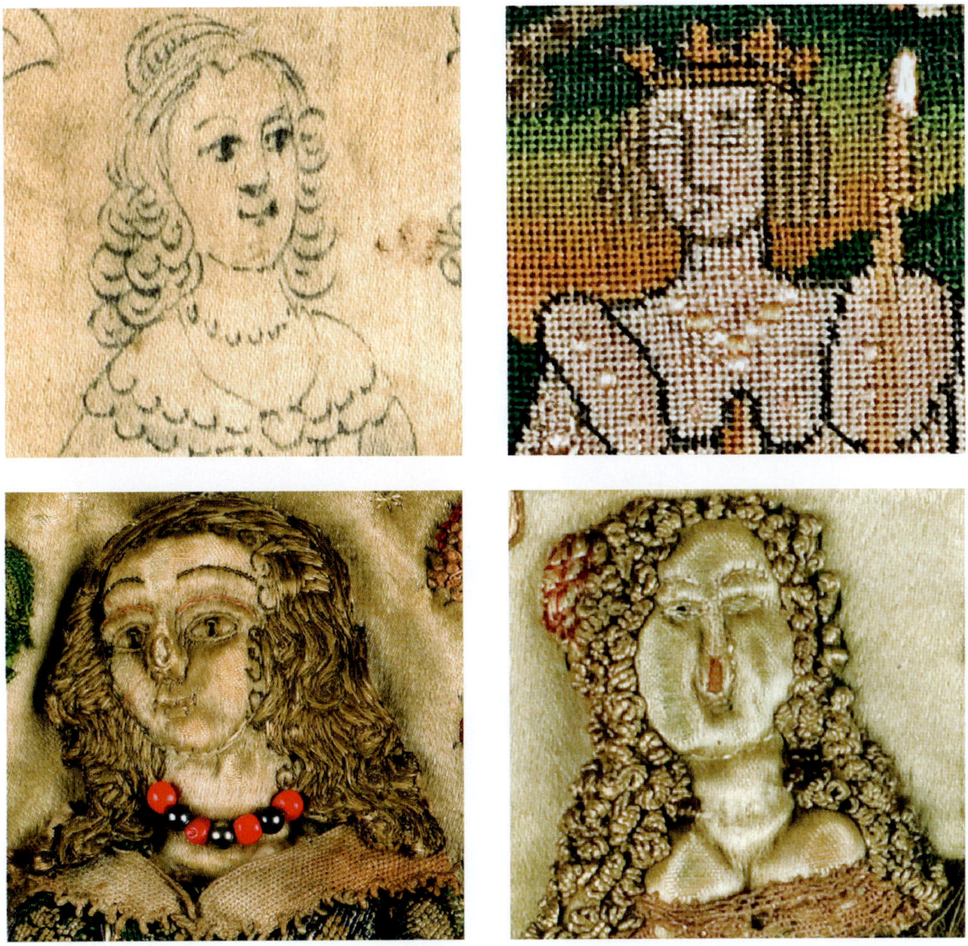

18 Faces and décolletages (clockwise from top left):
Pattern drawer's inked drawing, unfinished design (Feller Collection);
Queen of Sheba (detail, panel, Feller Collection);
Personification of the Sense of Smell (detail, mirror, Feller Collection);
Fashionable lady (detail, cushion, Feller Collection)

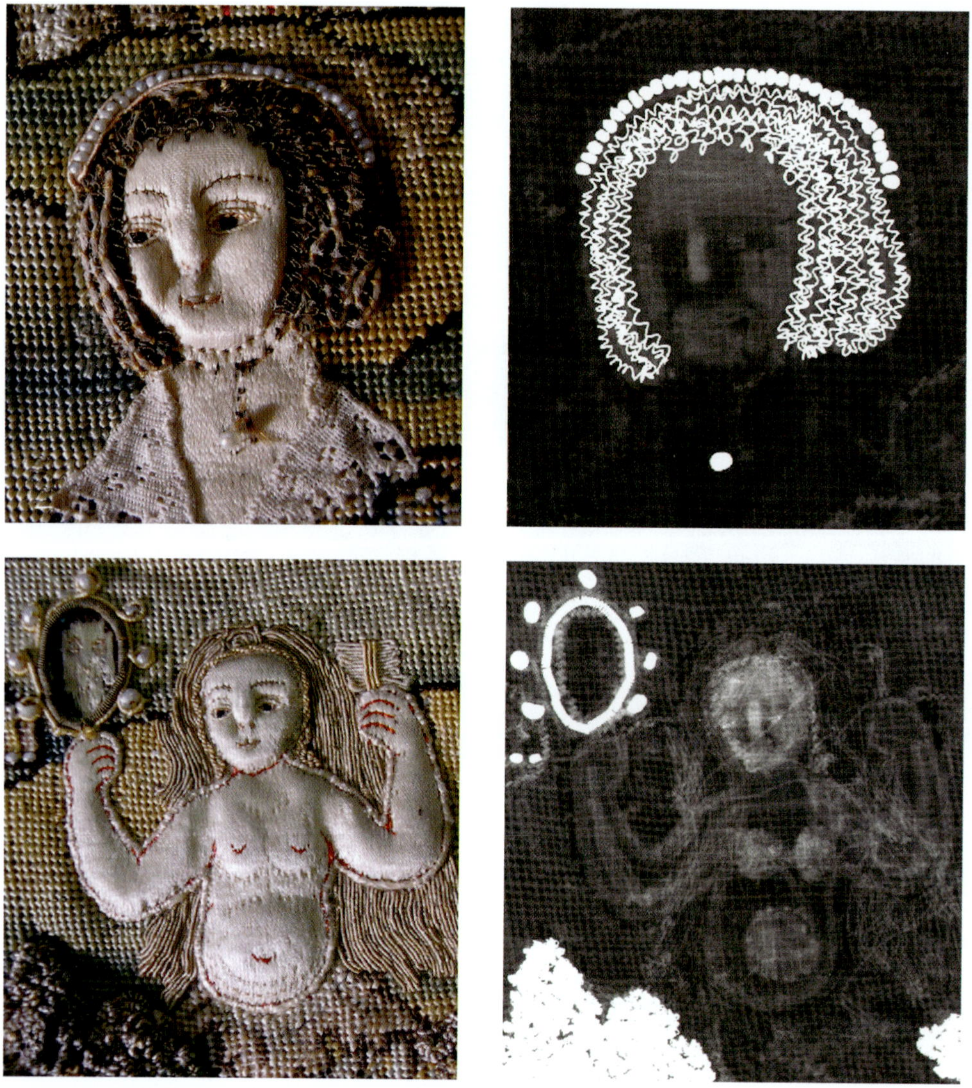

19 Solomon and the Queen of Sheba (clockwise from top left): Queen of Sheba, detail of embroidery; Queen of Sheba, radiograph; Mermaid, radiograph; Mermaid, detail of embroidery (Ashmolean Museum)

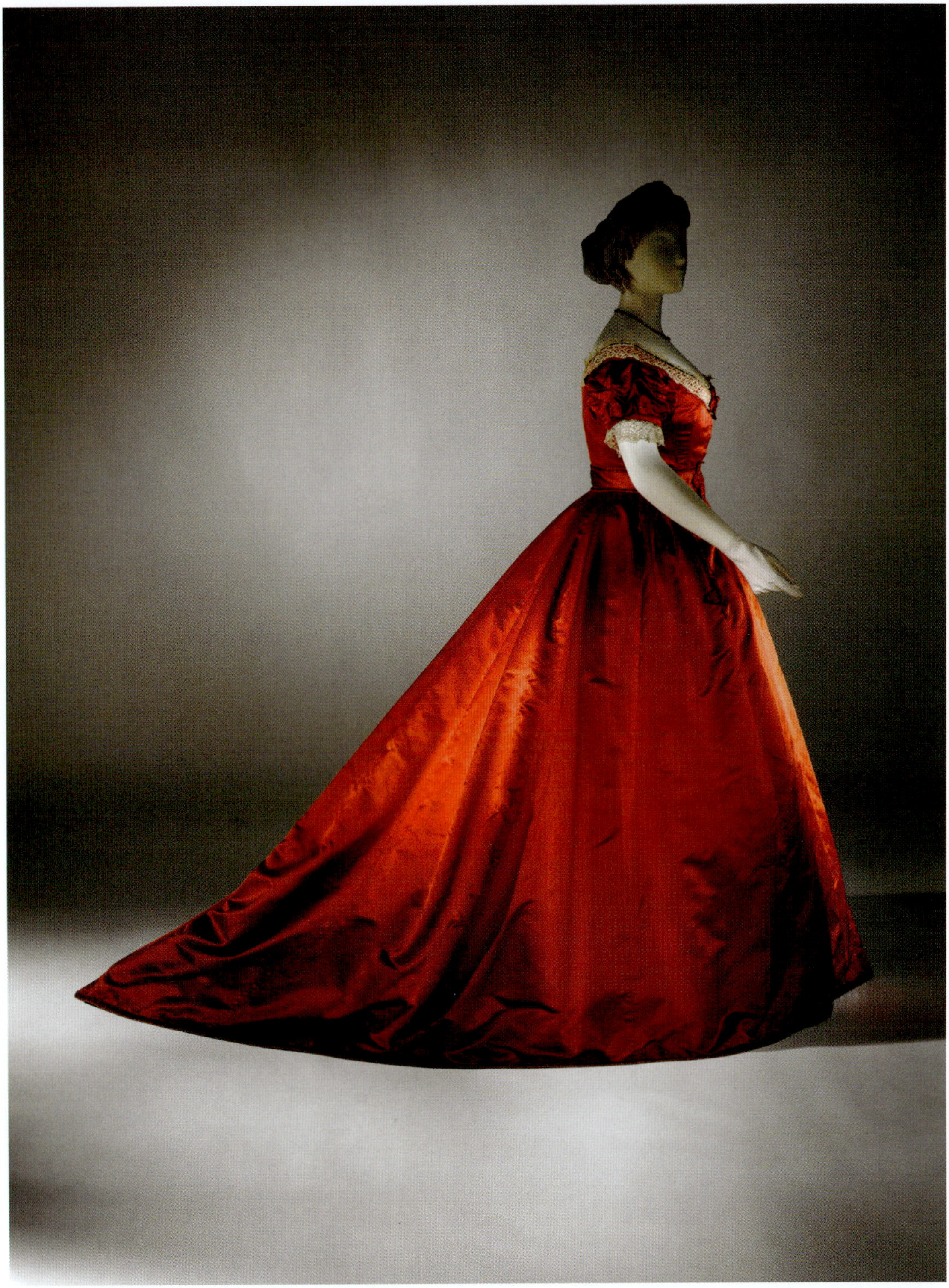

20 Evening dress, England, c. 1865, silk.

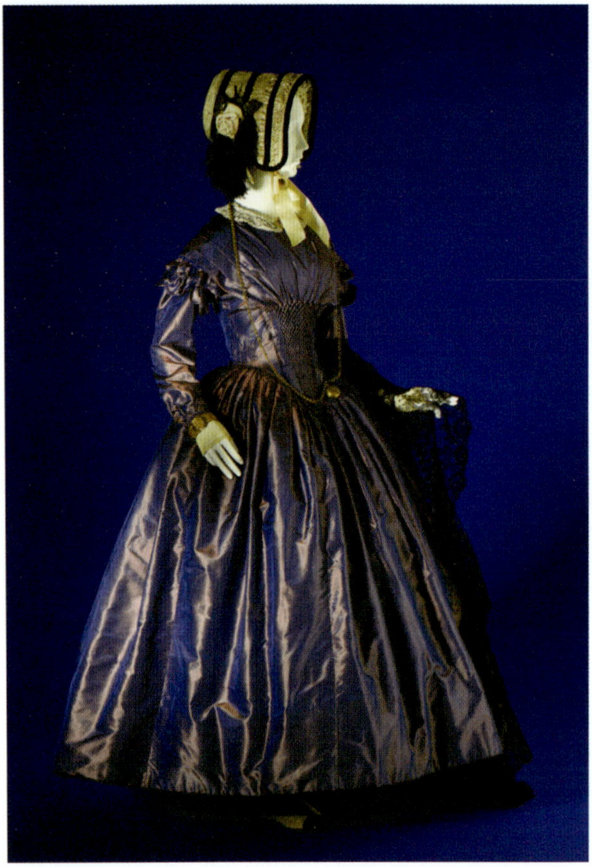

21 Day dress, probably France, c. 1845, silk taffeta

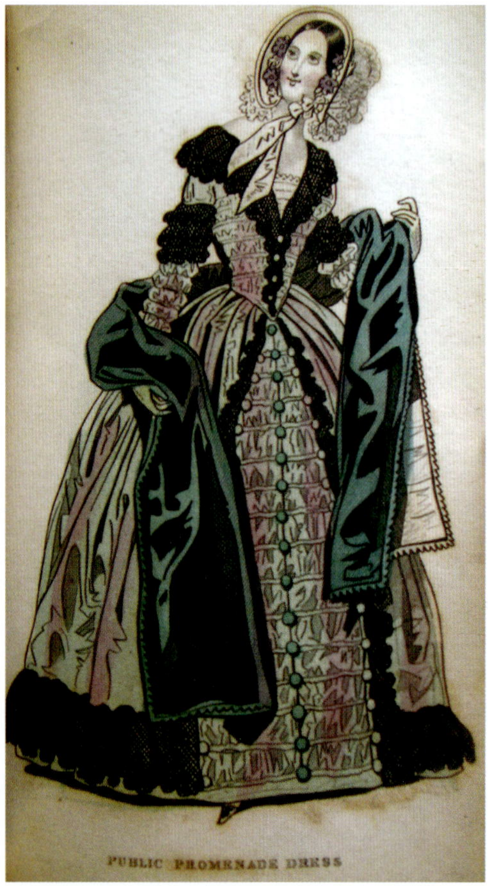

22 Fashion plate, 'Public Promenade Dress', *The Ladies' Cabinet* NS 6 (May 1841), watercolour and engraving on paper

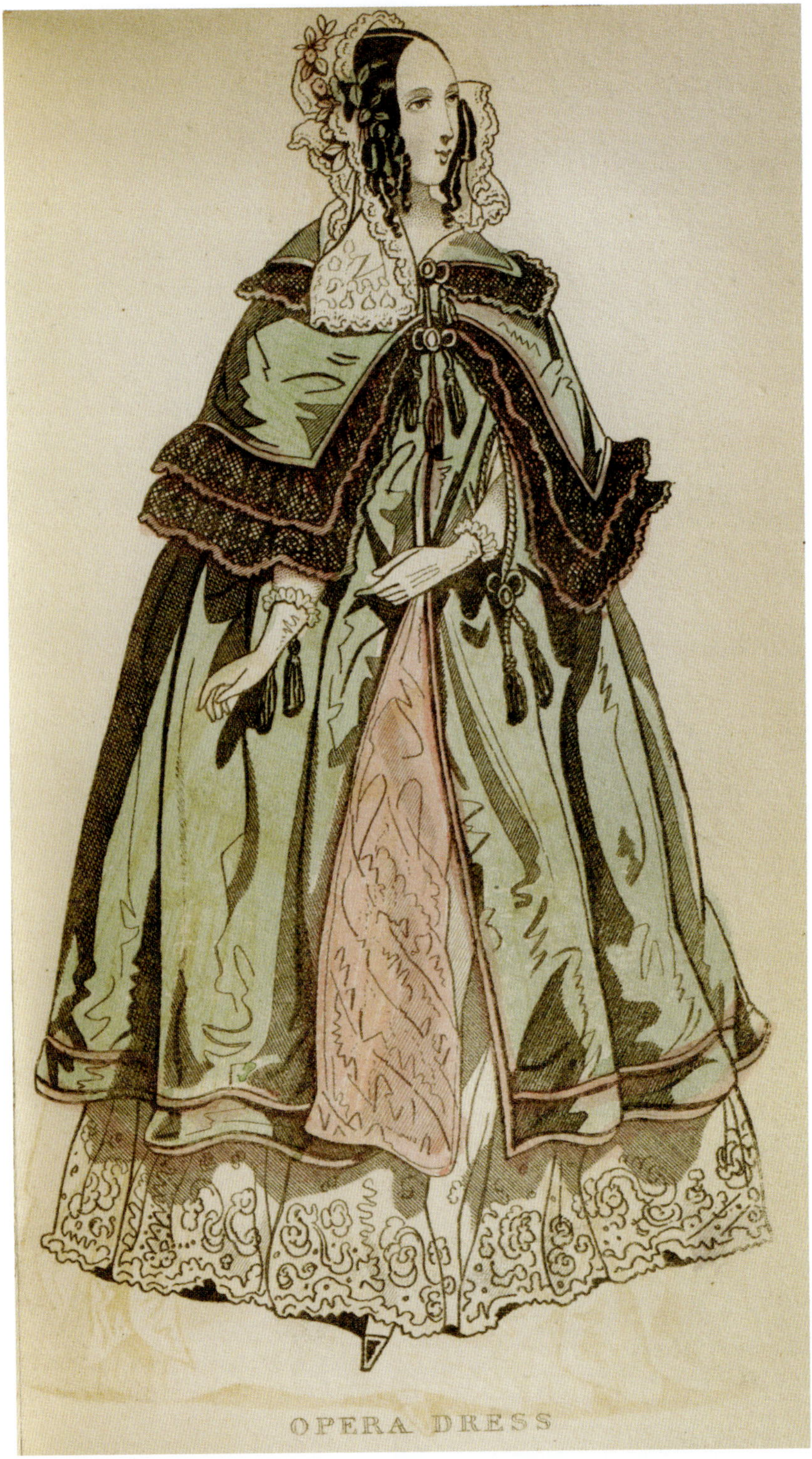

23 Fashion plate, 'Opera Dress', *The Ladies' Cabinet* NS. 7 (February 1842), watercolour and engraving on paper

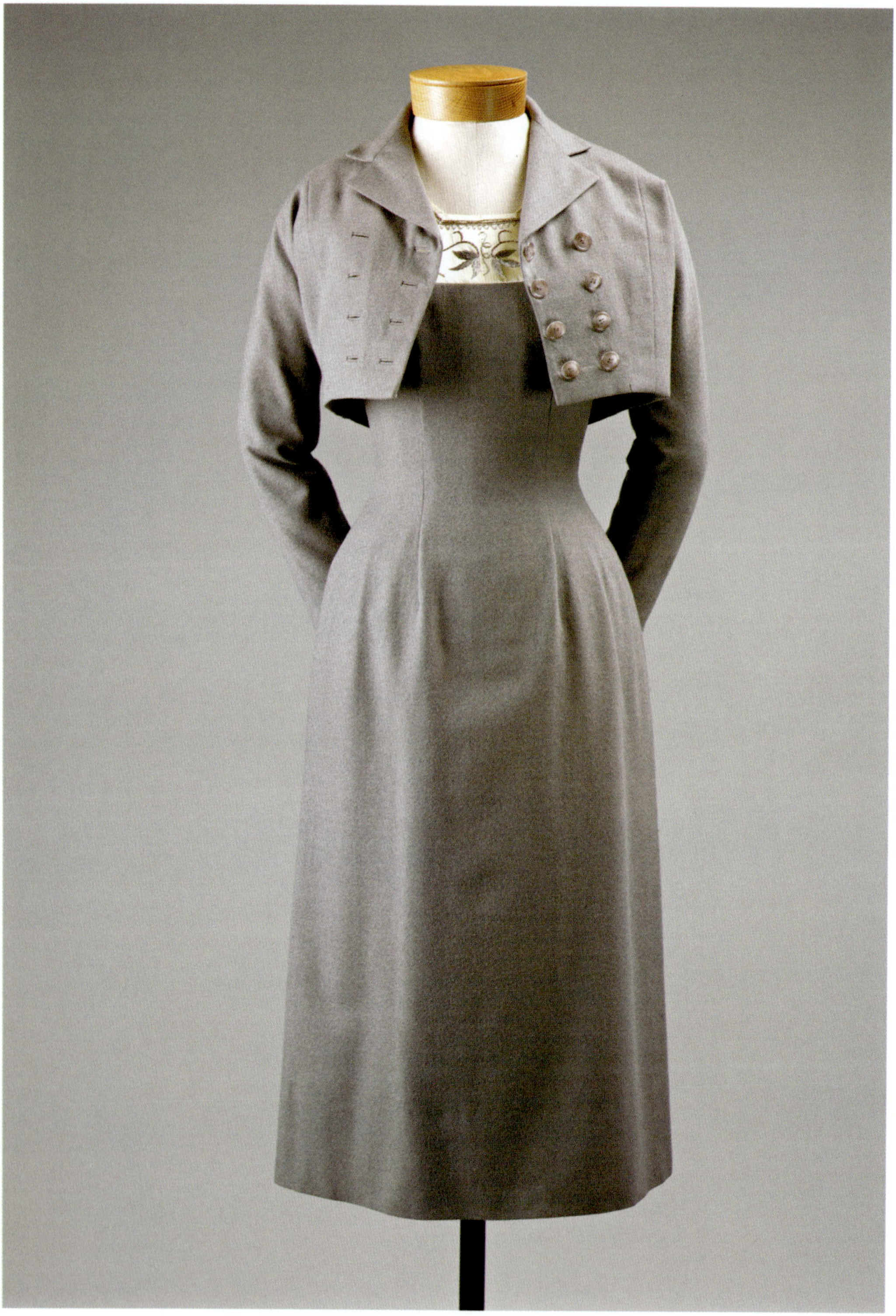

24 Anne Fogarty, Ensemble, fall/winter 1955–56

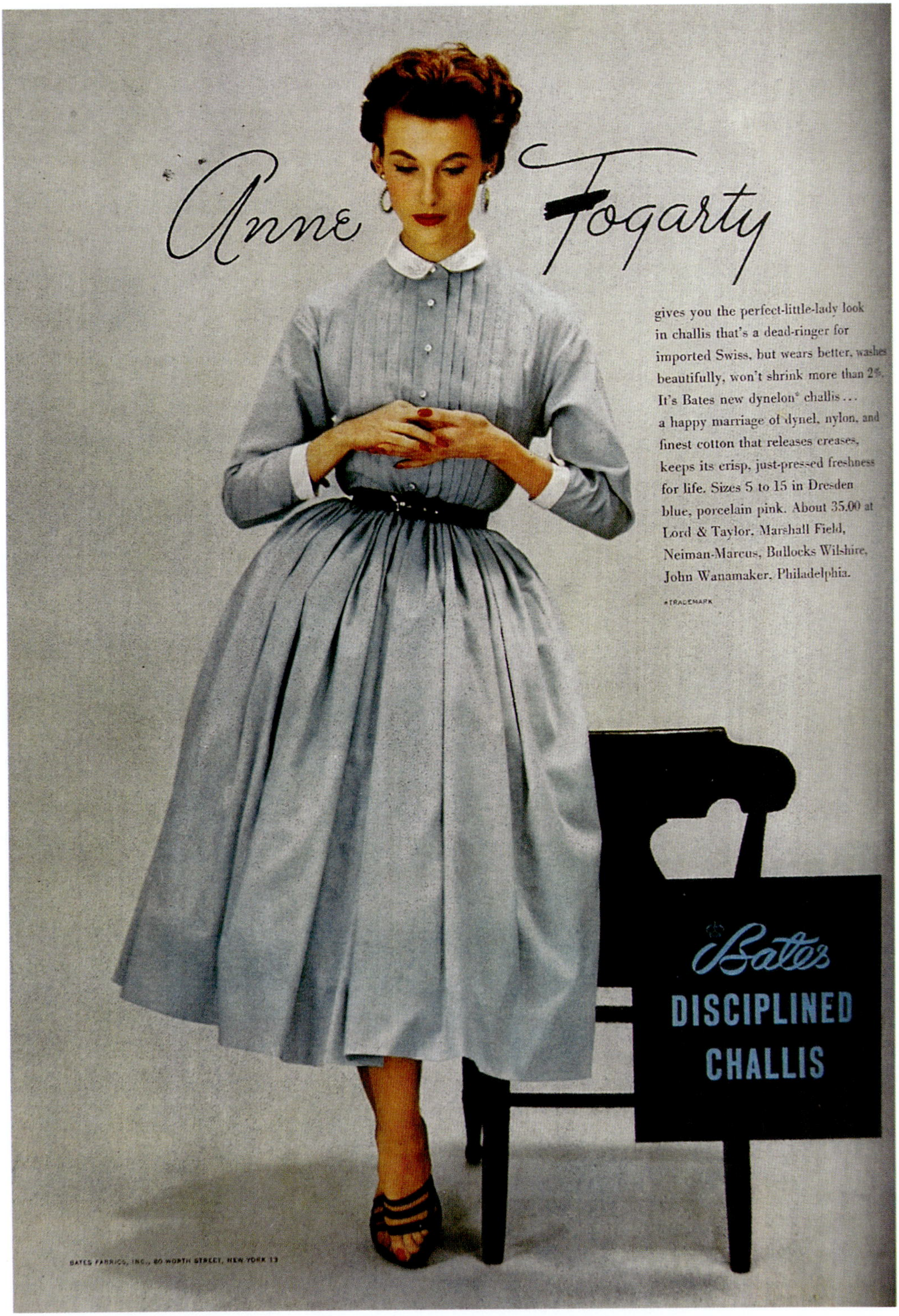

25 Anne Fogarty, Advertisement, 1953

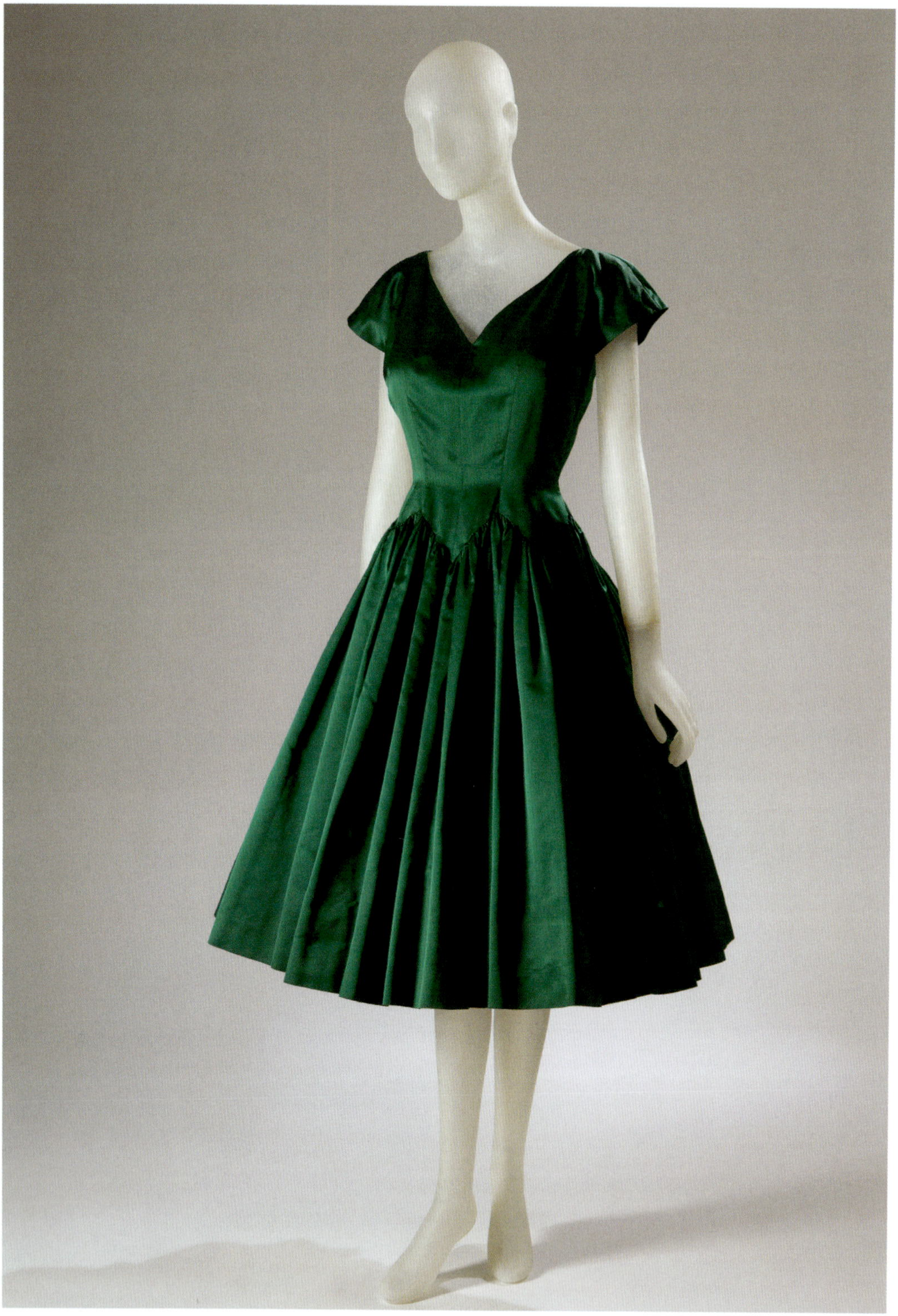

26 Anne Fogarty, Dress, 1950–54

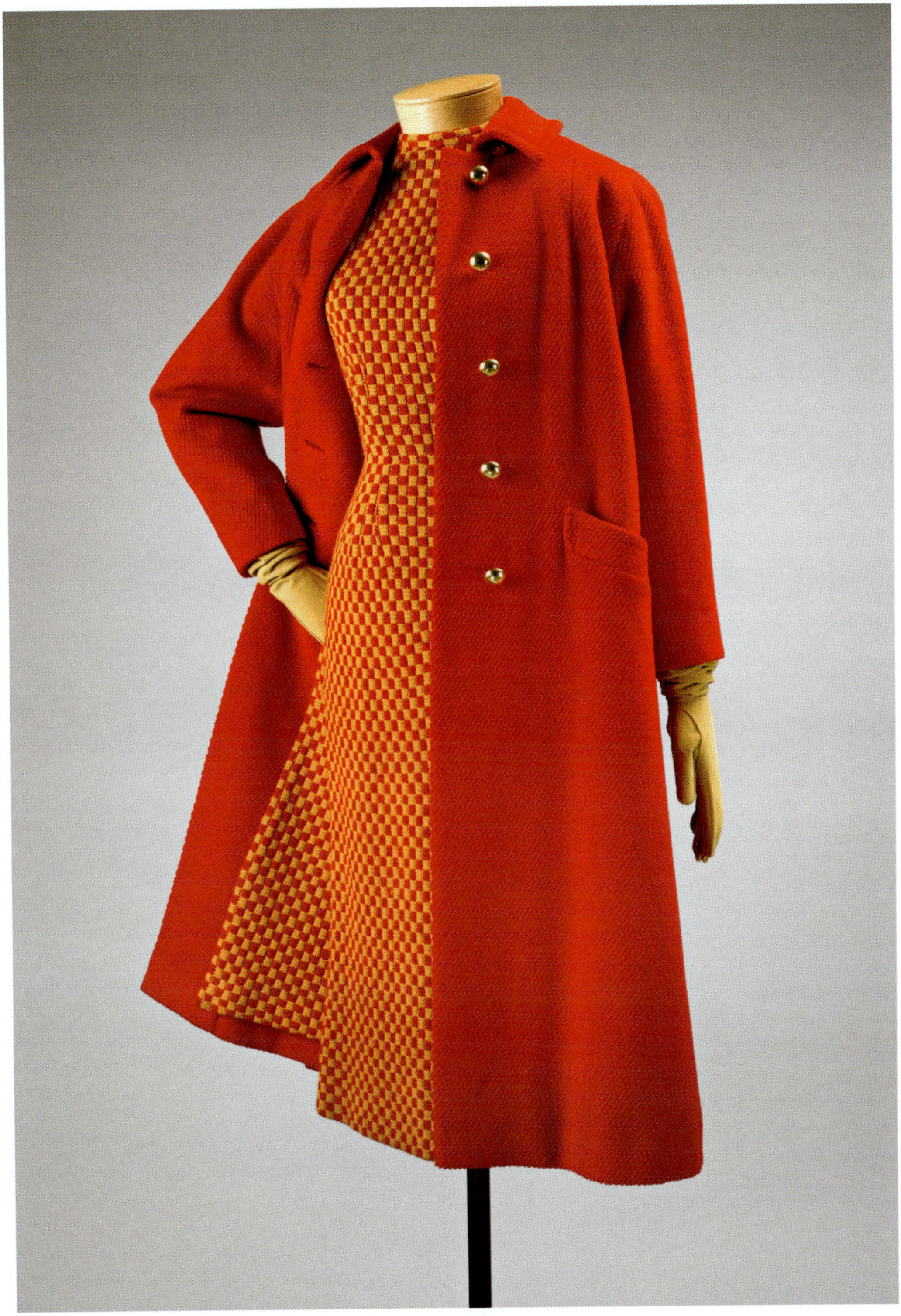

27 Anne Fogarty, Ensemble, fall/winter 1957–58

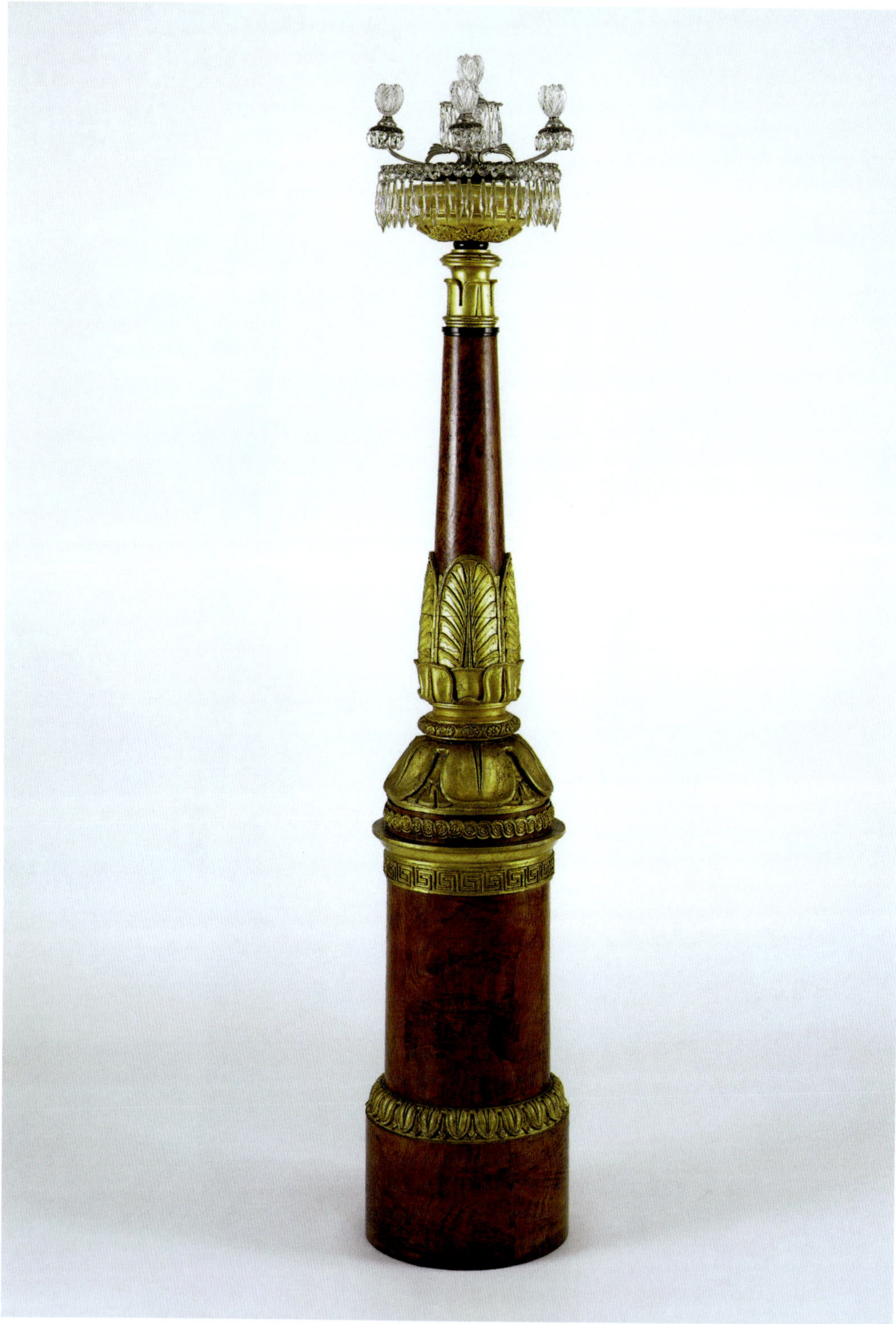

28 Torchère candelabrum, George Bullock, 1816–18

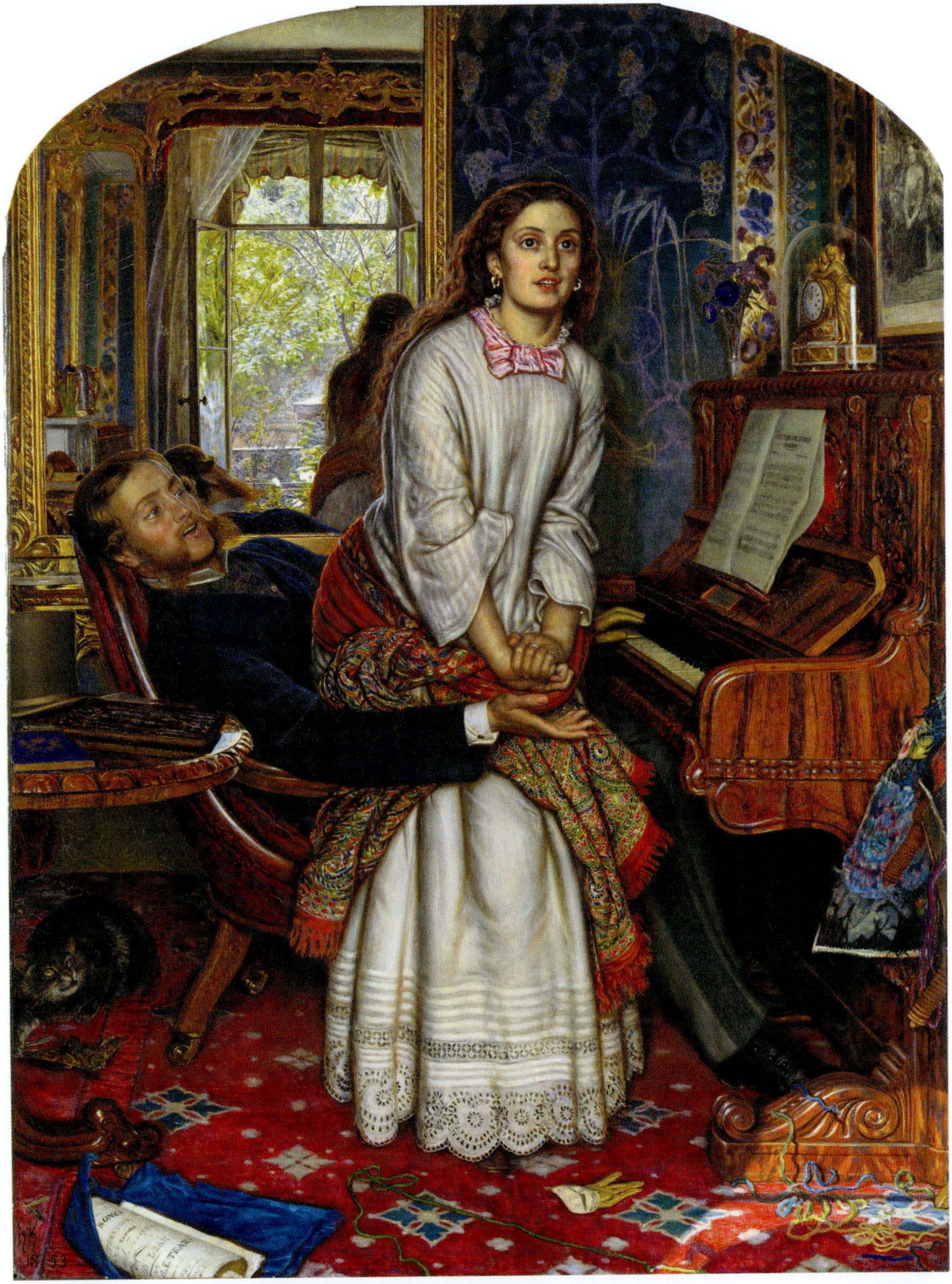

29 *The Awakening Conscience*, 1853, William Holman Hunt

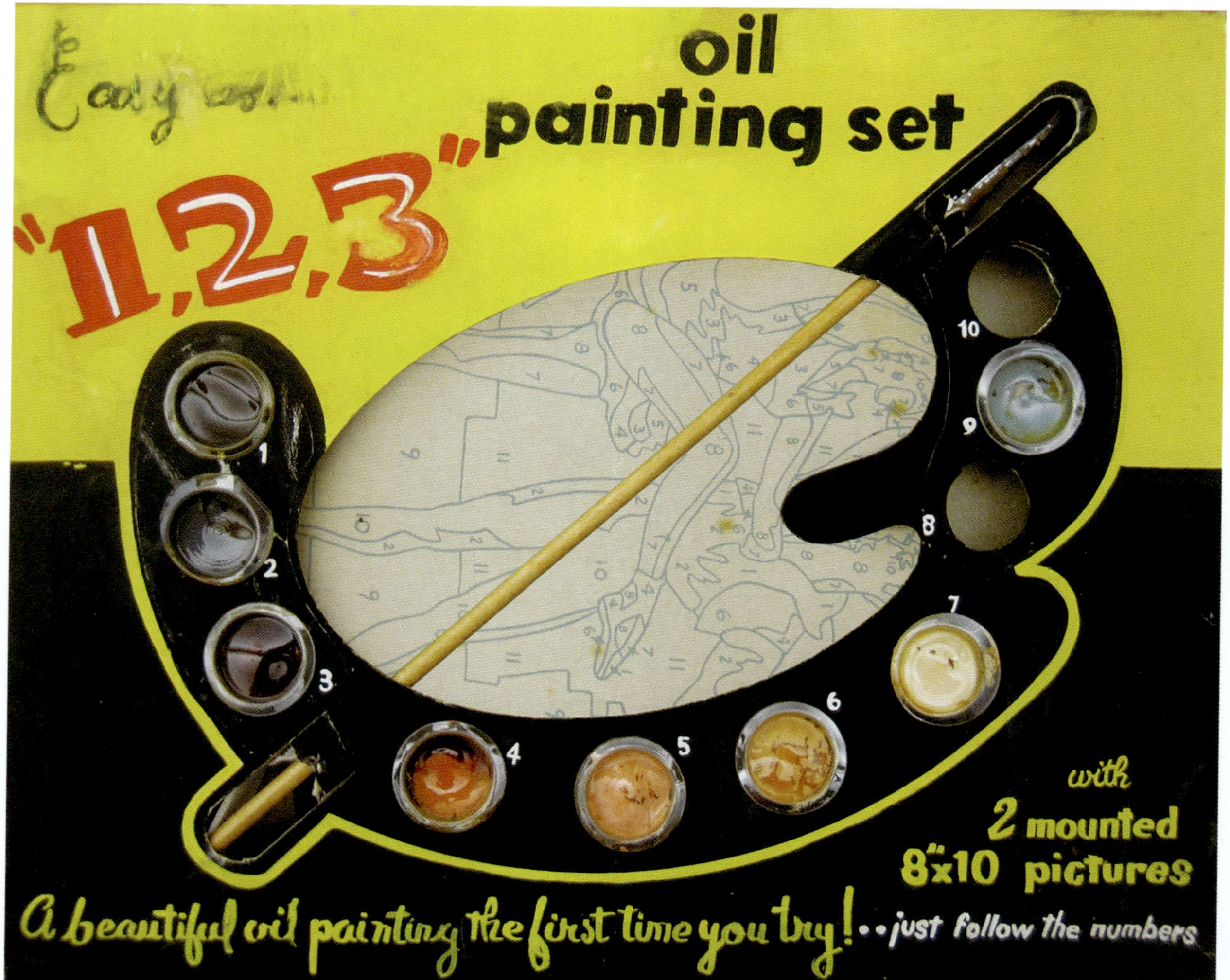

30 *Ballet Dancers "1,2,3"*. Oil painting set, Craft Master Palmer Paint Inc., 1953

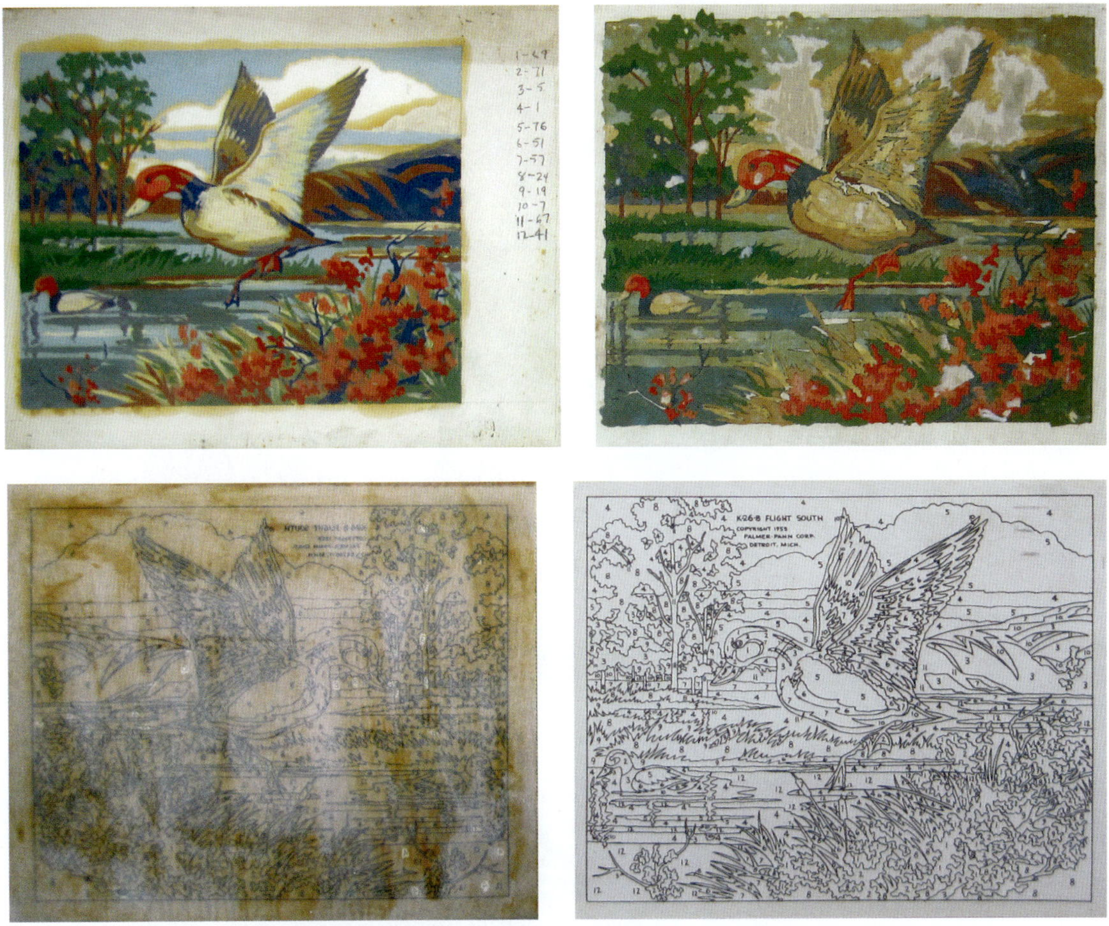

31 *Flight South*. Preparation sketches, samples and test runs, Palmer Pann Corp., 1955

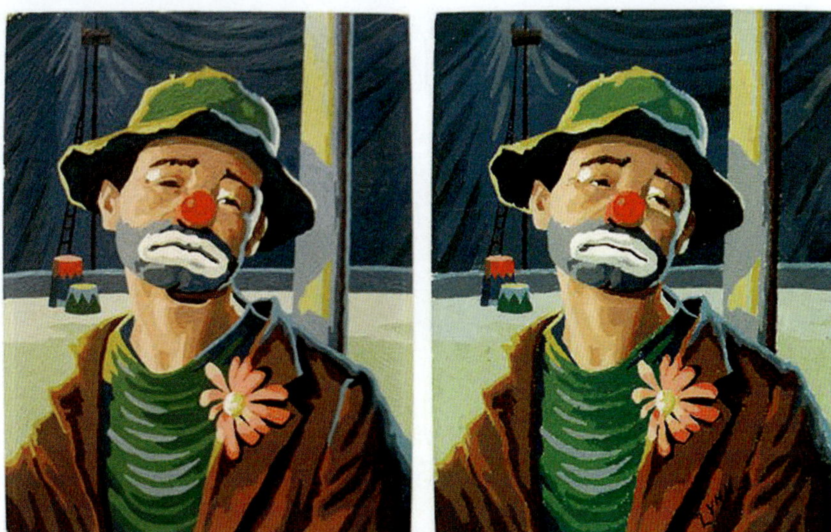

32 *Old Sadface*. Craft Master kit, 1969

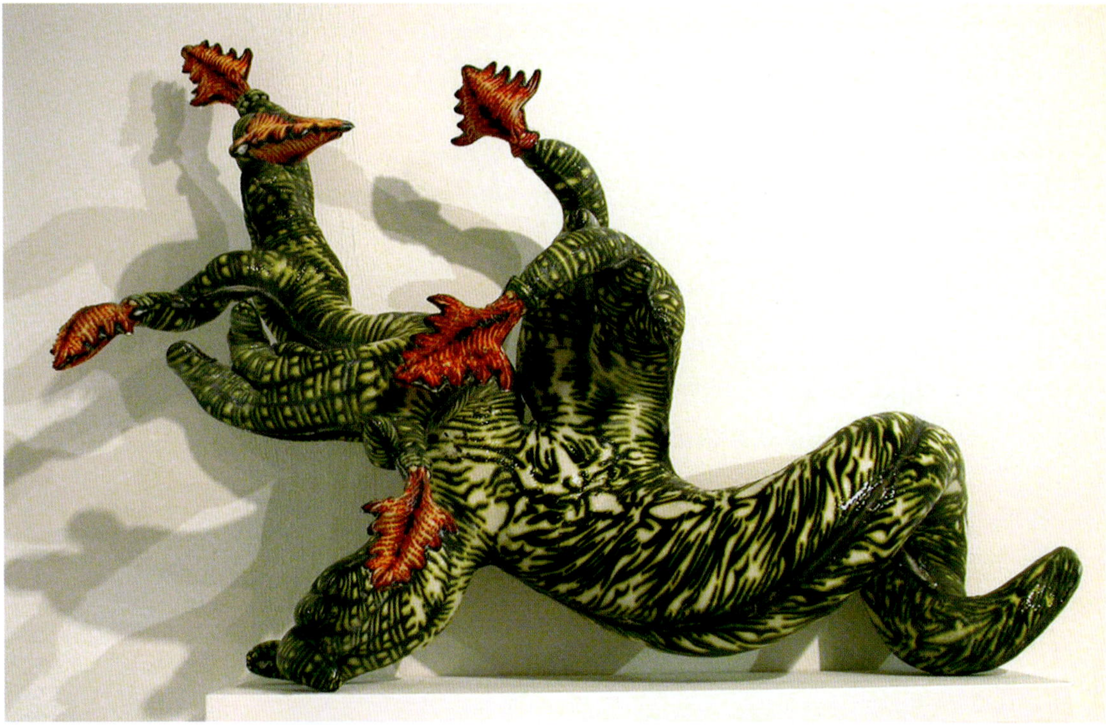

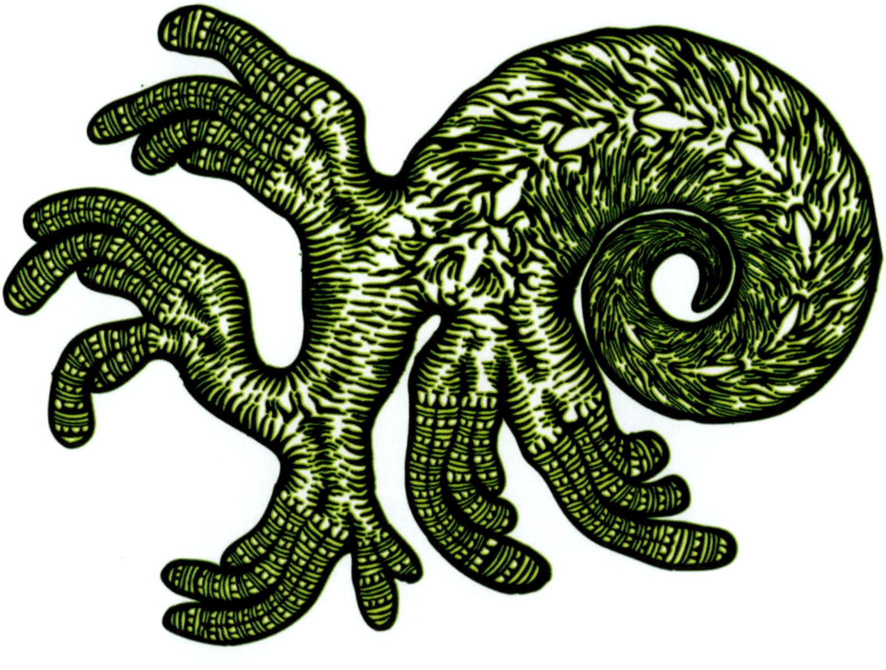

33 *Mandrake,* 2009, porcelain with screenprinted in-mould decoration

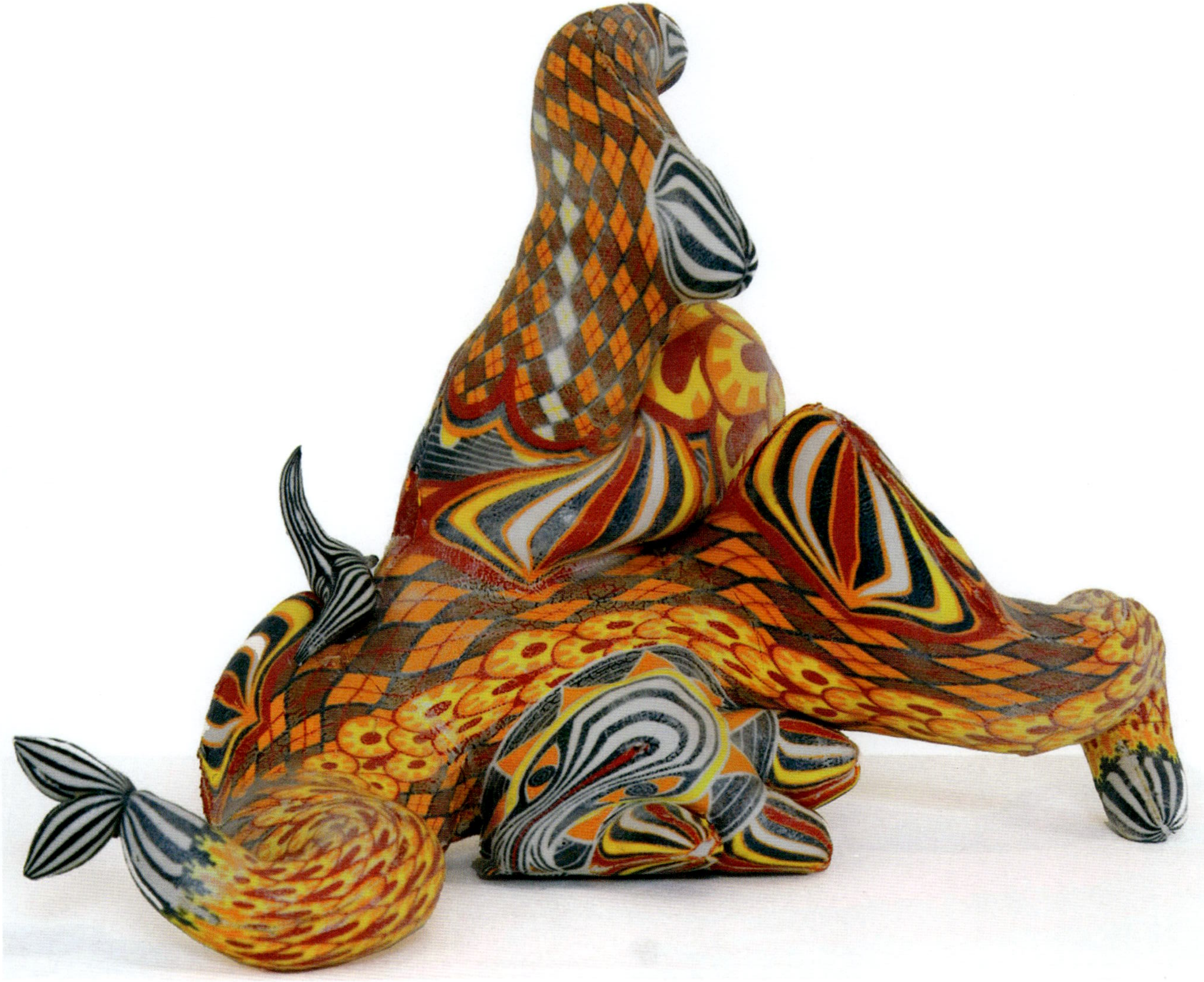

34 *Arlecchino the Brave*, 2009, porcelain with screenprinted in-mould decoration

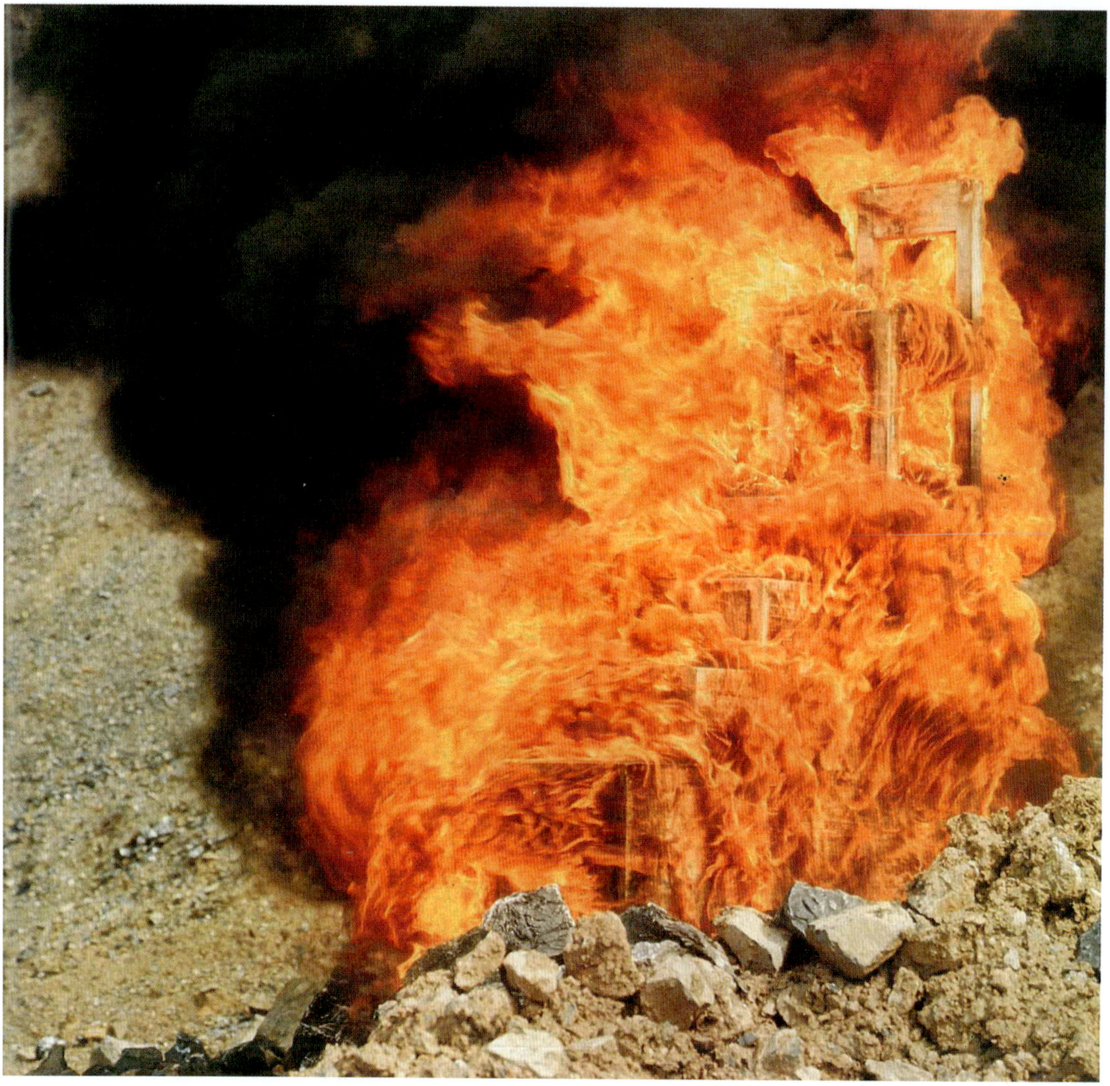

35 Alessandro Mendini, burning of Lassù chair, for *Casabella* magazine, 1974

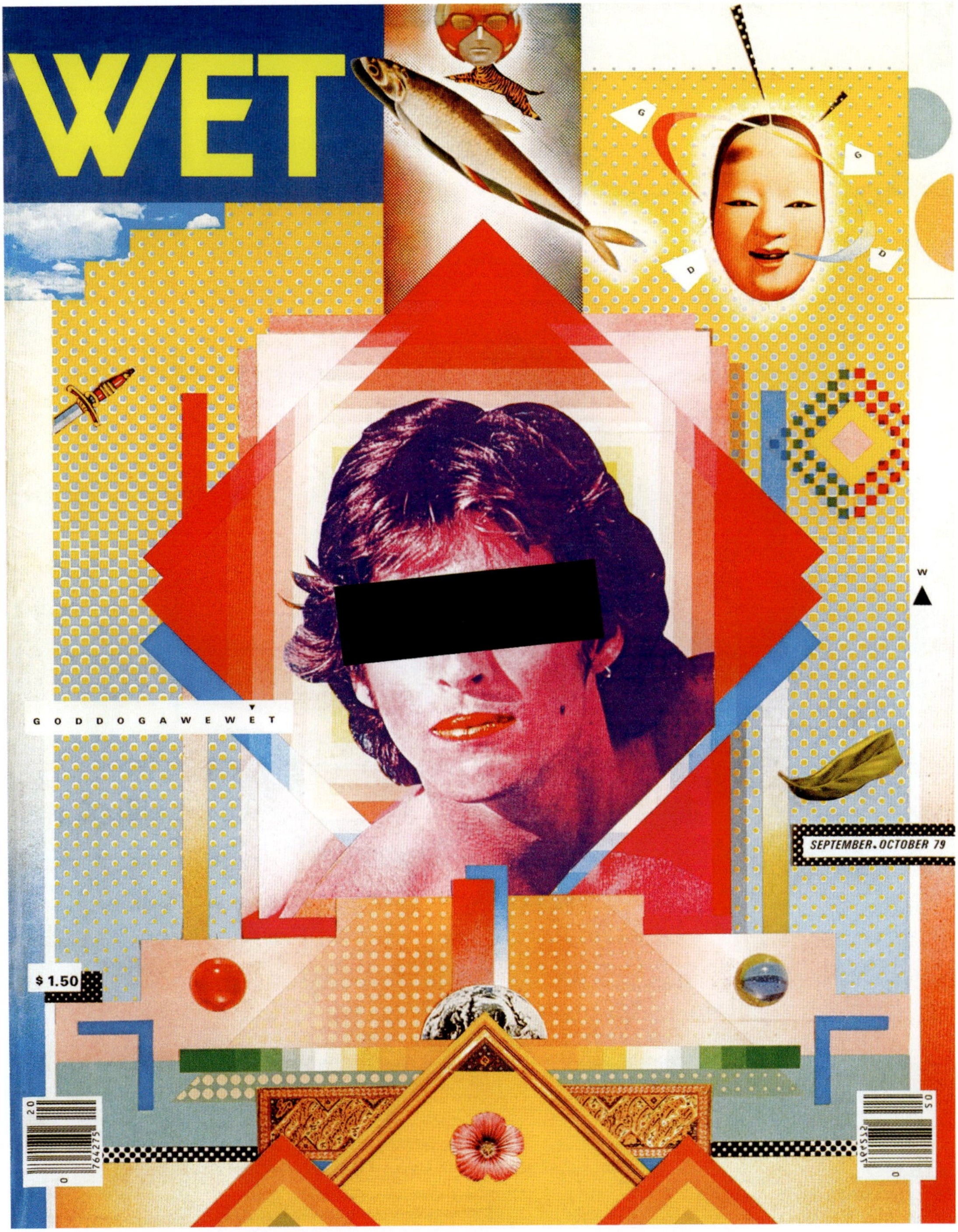

36 April Greiman in collaboration with Jayme Odgers, cover for *WET* magazine, 1979

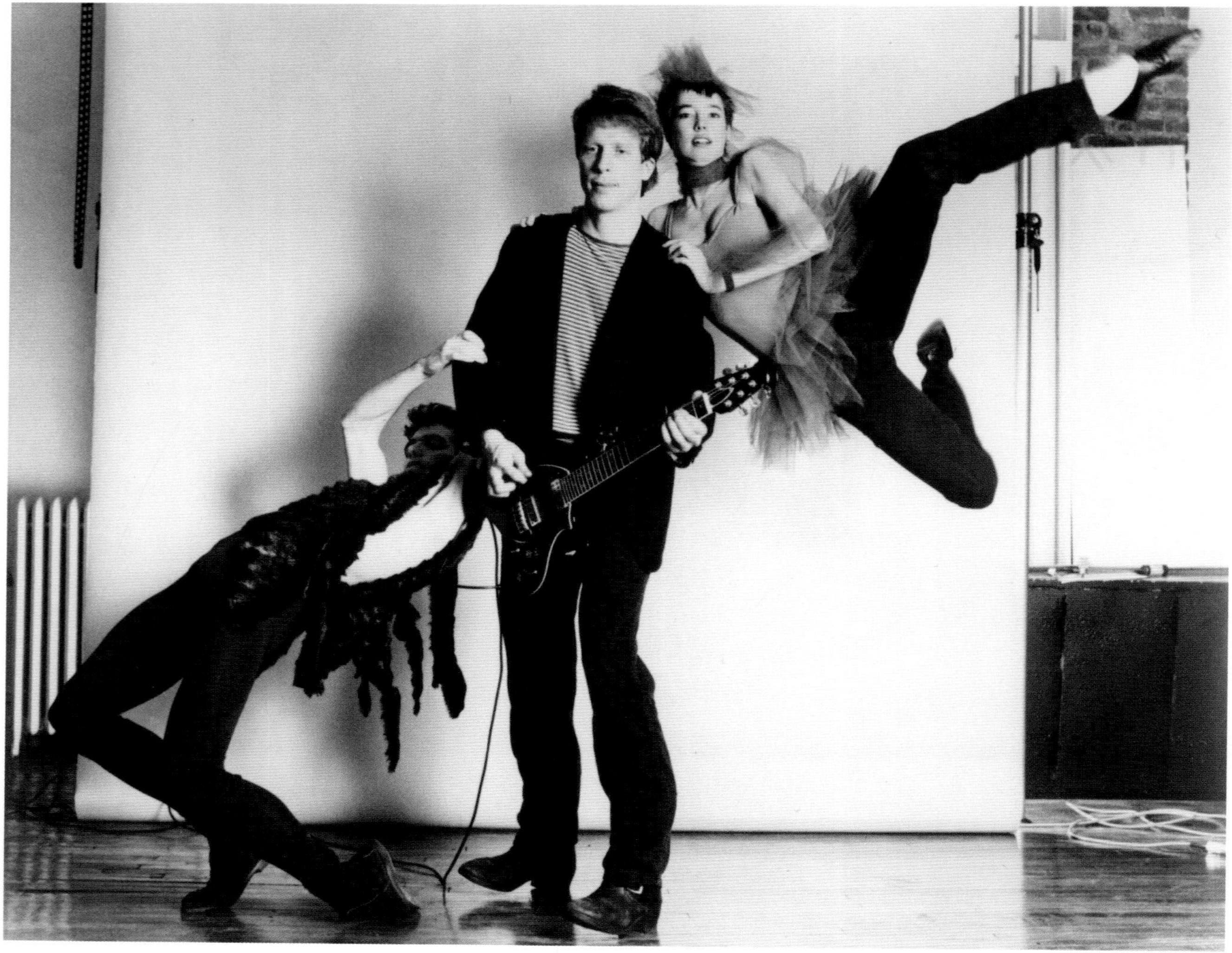

37 Karole Armitage (choreographer), *Drastic Classicism*, 1981. Left to right: Joseph Lennon, Rhys Chatham, Armitage. Costumes by Charles Atlas

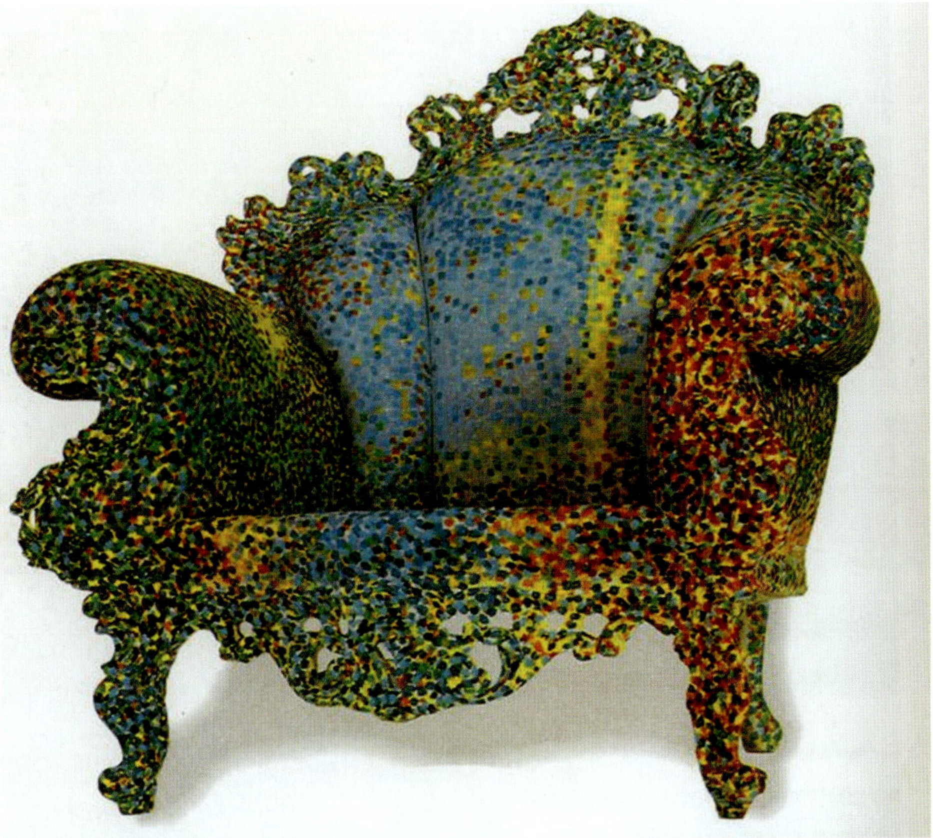

38 Alessandro Mendini, *Proust Chair*, 1978. Wood and painted textile

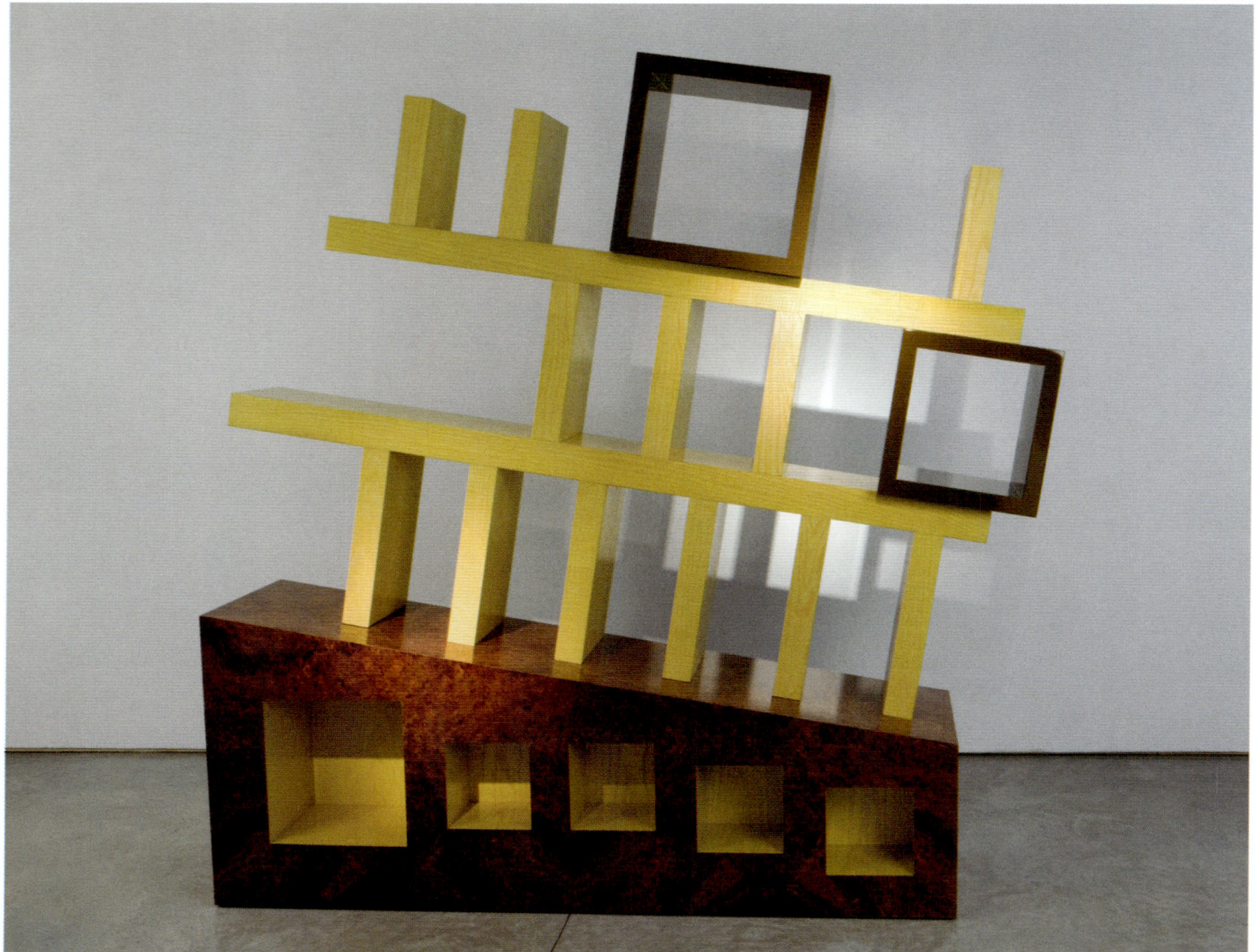

39 Ettore Sottsass, Cabinet No. 8, 1994. Yellow-lacquered maple wood with waterstain grain, natural Burled Elm. Edition of 6

the electricity of our bodies … and it heightens colors with a charming light. It possesses a cheerfulness of which the dull surfaces of wool and linen are destitute.'[16] The shininess of silk intensified colours, a quality that can be observed in surviving examples of women's dresses coloured with vivid dyes (Plate 20). Shoeser emphasizes that enthusiasm for brightly coloured silks reached a peak in the nineteenth century not since surpassed.[17]

Contemporaries, however, did not only discuss brightly coloured textile surfaces. In the *Englishwoman's Domestic Magazine* in February 1875, the fashion writer reported to her readers about poplins, in this case woven of silk and wool. Praising the colours, the author also commented favourably on the quality of the shades of black achieved in some of these textiles: 'Their poplins are so soft in texture and so beautiful in colour that they appear as though they had issued from some fairy loom … I saw there a black poplin, the best shade of *real* black I ever saw, with a more silky, sheeny surface than a silk of almost double the price.'[18] In the eyes of this writer, this poplin was remarkable for both the quality of the black shade *and* its glossy surface, properties that lent a seemingly magical quality to these textiles.

The shot, or 'changeable', silks popular in the 1840s (Plate 21) required especially fanciful phrases from fashion writers. Reports in the *Ladies' Cabinet* noted a dress silk that had 'a ground resembling mother-of-pearl, shaded in the colours of the rainbow' and 'a shot silk of the colours called *gorge de Pigeon* [pigeon-throat]' for a morning visiting dress.[19] In May 1843, an article in the same magazine drew attention to the continuing popularity of the '*mantelet cameleon*', describing one in a 'novel and pretty' material: 'it is a changeable silk, which, under three different points of view, gives three distinct colours, and these tints are extremely rich and brilliant'.[20] The visual effect of this textile's surface was also noted, in similar language, in the *Lady's Book* a year later, in June 1844. The fashion writer included a report from a Boston newspaper which noted 'For dresses, chameleon silks are much worn – three distinct colours, by some magic of art, being thrown on a plain ground, looking as if "Iris dipt the woof"'.[21] Some contemporary visual representations of shot silk garments attempted to show their changeable surfaces by using two different colours (Plate 22). The various colours that could be seen from different vantage points were a significant and aesthetically attractive aspect of these textiles to contemporary commentators, ascribed in one case, as noted above, to a kind of magic. These descriptions also exemplify contemporary interest in the difference that multiple perspectives made when viewing an object.[22]

A humorous poem titled 'Bonnets', published in the *Lady's Book* in 1840, also employed this kind of language to discuss the appearance of textiles used in fashionable headwear:

Dizzily your tints perplex us;
Satin, velvet, leghorn, crapes,
Rise, like Hamlet's ghost, to vex us,
With your strange, unearthly shapes.
As the rainbow's colours, various
Are the hues which ye display;
And chameleon-like precarious,
Changing one by one away.[23]

Like a fashion writer, this author called upon readers' imaginations and emphasized the dizzying variety of colours available by referring to the chameleon and the rainbow, underscoring contemporary awareness of the unstable nature of vision. The eye could be bewitched, especially when looking at colour. These aspects of colour and its fashionability were mined for humour, demonstrating how these notions were securely established aspects of the commentary on colour and its perception. All the 'chameleon' references discussed above illustrate the mid-nineteenth-century attention to and pleasure in the interaction of colour and light on these visually remarkable textile surfaces.

Science, vision and colour

Another chameleon appeared in the *Lady's Book* in 1854, outside the fashion reports. A recurring feature, called 'Chemistry for youth', contained directions about how to make a 'Mineral chameleon'. The reader was instructed to make a powder of oxide of manganese and nitrate of potass and was then told: 'If a small quantity of the powder be put into a glass of water, it soon forms a green solution, when it passes into a violet hue, and lastly it becomes of a beautiful red color … The changes of color that occur are owing to the different degrees of oxidizement of the metal.'[24] The inclusion of this piece in a woman's magazine reveals how women were encouraged to participate in amateur science in the nineteenth century as part of their social and cultural duties, as well as highlighting the visually entertaining nature of some of this science.[25] The account of this 'chameleon' also illustrates the contemporary interest in the observation of changing colours, echoing the descriptions of shot silks.

Other eye-catching engagements with science were also cited in discussions of dress textiles and fashion. The author of a September 1854 article in the *Lady's Book* about the fashionable textile *moire antique* referred to the 1815 invention of the kaleidoscope by David Brewster (1781–1868).[26] Matilda Browne described how, while in a London shop, newly-arrived textiles 'were spread out for my inspection, and the table soon looked like a kaleidoscope, the colours were so brilliant, so exquisite in hue'.[27] For many contemporaries, the shifting colours and images of the kaleidoscope represented modernity.[28] Isobel Armstrong notes that the kaleidoscope was 'one of the earliest specifically nineteenth-century optical crazes for popular media', highlighting it as an example of the overlapping of entertainment and scientific research.[29] Helen Groth stresses that the kaleidoscope enabled 'the visual transformation of the incidental aspects of everyday life into objects of interest'.[30] These periodical references to the kaleidoscope in discussions of dress textiles thus emphasize their variety and modernity, showing how they provided yet another visually arresting way to experience colour in daily life.

Related to popular interest in the field of optics in nineteenth-century magazines were references to contemporary colour theory. Working in a tradition of commentary that stretched back to Isaac Newton (1643–1727) and Johann Wolfgang von Goethe (1749–1832), French colour theorist and Gobelins dyemaster Michel-Eugène Chevreul (1786–1889) was one of the greatest influences on mid – to late nineteenth-century ideas about colour.[31] Chevreul wrote several books outlining his theories in the mid-nineteenth century and his fame derived largely from the way in which he codified and organized colours.[32]

Chevreul addressed clothing, offering advice about colour combinations explicitly to portrait painters and designers of military uniforms, but also hoping that 'many classes of artists, particularly dressmakers … will derive some benefit from consulting them'.[33] His interpreters quickly adapted his suggestions as general fashion precepts for contemporary female dress. Much of this advice hinged on Chevreul's law of simultaneous contrast, the effects of what are now called afterimages. Placed next to each other, colours appear differently to the eye than they do when viewed alone. Both men of science, such as Brewster, and philosophers, such as Goethe, enthusiastically studied afterimages in the late eighteenth and early nineteenth centuries.[34]

The discussions of these retinal effects that permeated fashion advice in mid-nineteenth-century women's magazines occasionally mentioned Chevreul by name; he was known as a public lecturer on colour.[35] More frequently, however, the articles contained long lists of colours that were supposed to flatter different complexions, based on Chevreul's published discussions of how colours interact. These pieces often suggested experimentation with slips of coloured paper or pieces of coloured ribbon to demonstrate (in the words of Charles Martel, one of Chevreul's English translators) the 'optical composition' of his law of simultaneous contrast.[36]

These experiments are related to Isobel Armstrong's discussion of nineteenth-century re-creations of Isaac Newton's famous division of light into the colour spectrum using a prism. She argues that Newton was especially significant because, through him, 'the *household shutter*, the darkened *domestic* room, can be made to bring the prismatic radiance of the spectrum and the splitting of light into the core space of the modern subject'.[37] The mineral chameleon, the kaleidoscope, the slips of paper and the prism all demonstrate how middle-class women were encouraged not only to teach themselves about science and colour, but to notice and experience personally everyday applications of science in domestic spaces.

Matilda Browne explicitly referred to light in her March 1872 fashion report: 'It is especially with regard to the effect light produces upon them that materials differ one from another. Each receives the luminous rays in its own fashion: some, like wool, completely absorb them; others, like satin, reflect them brightly; some dull them, as cloth does, and some completely darken them, as velvet.'[38] This observation of how different textile surfaces reflect or absorb light is another example of the nineteenth-century attention to vision, showing popular interest in scientific enquiry into the behaviour of light. By perceiving the ways in which scientific phenomena appeared in daily experiences, including the seeing and wearing of clothing, women could display their proper understanding of the world.

Clothing in different kinds of light

The final intersection of colour, fashion and vision explored in this chapter is the mid-nineteenth-century discussion of textile colours and different kinds of light (natural or artificial), as well as fashionable women's awareness of these variations. The adoption of new lighting technologies, such as gas light, was uneven and in some cases resisted. By the mid-nineteenth century, however, enough interior spaces were lit by gas to warrant relevant colour advice, especially for evening dress.[39] These references to different qualities of light illustrate one way in which women engaged with technology through fashion and dress.

Women's magazines contained references to shops that provided customers with gas lit spaces in which to choose dress textiles. Charles Frederick Worth (1826–95) opened his haute couture house with Otto Bobergh in Paris in 1857–58 and as Diana de Marly mentions, his establishment included a *salon de lumière* 'lit by gas ... with all daylight excluded'.[40] Almost a decade earlier, however, in October 1849, the *Lady's Book* had reported on recent improvements to Levy's department store in Philadelphia. The reporter noted, 'one great thing to be commended, is the room where ladies can see the effect of silks by gas light. How often the complaint is made, "It looked beautiful when I bought it; but it looks entirely different at night." This can now be obviated; for ladies will have an opportunity of judging the materials they purchase by both day and gas lights.'[41]

References to the appearance of different textiles in candlelight, gas light and occasionally daylight, pepper the fashion reports of the magazines examined. In some cases, light was explicitly included in colour names. Reports during the 1840s in the *Ladies' Cabinet* mentioned a colour called *vert lumière* (Plate 23), described as 'a beautiful bright shade of light green' and the *Englishwoman's Domestic Magazine* reported a 'new gas-green colour' in March 1868.[42]

More common than explicit names, however, were comments about the appearance of coloured textile surfaces in certain light conditions. In December 1847, a writer for the *Ladies' Cabinet* described evening dresses 'composed of *brocatelles* and damasks, with dead grounds, figured or striped in satin'. The article went on to observe, 'The effect of these materials is very beautiful by candle-light.'[43] Two decades later, in September 1867, Matilda Browne noted that the season's novelty was 'shot silk', listing several colour combinations: 'amongst the favourite tints for dress to wear in the daytime, and for the evening there is quite a series of soft colours, blue and white, mauve, green and rose-colour and white – very brilliant and remarkably pretty by gaslight'.[44]

Sometimes a writer singled out the appearance of a specific colour, as when the monthly fashion report in the *Lady's Book* of November 1869 commented that *paon*, or peacock colour, was appropriate for carriage or walking dress, but that 'it doesn't look well at night'.[45] In February 1867, Matilda Browne noted the fashionability of 'the Regina violet' described as 'a light bright tint, more inclining to red than blue'. The author went on to say that this colour 'is very fashionable this year for ball toilet[te]s, as it looks particularly well by candlelight, an advantage possessed by very few shades of this colour'.[46]

When discussing light conditions, as with many recommendations to middle-class readers about appropriate dress, writers evoked an ideal female consumer operating in an elite world of fashionable consumption. Matilda Browne noted in March 1872, 'Nor will a lady of taste forget that colours change according as they are looked at by day or by lamplight, and we see her in the middle of the day, stepping into a closed saloon, lit up with gas, to choose her evening dresses.'[47] Although shopkeepers' trade publications of the period did acknowledge the effect of gas light on the perception of coloured textiles, this shopping vignette may have been a fantasy for some readers.[48] The text, however, encouraged the readers' powers of imagination and underscored the everyday interactions with light and colour that women could experience through their dress.

The most specific information about which colours were particularly affected by gas light appeared in two articles by Matilda Browne in the *Englishwoman's Domestic Magazine* in March 1872 and November 1873. These were both quite detailed in their lists of colours either enhanced or harmed by gas light. Because of the yellow quality of gas light (and candle light) Browne and the writers of other such articles urged their readers to consider the amount of yellow in a dress textile for evening wear. In the 1873 article Browne observed, 'There are, for instance, yellowish greens, which appear gaudy and crude in the daylight, but which become delicately beautiful when the effect of gaslight has taken from them the yellow glare. Blues of an intense turquoise hue look best at night; yellows should be very *prononcé* to show at all for evening dress.'[49]

Browne provided very similar examples in the earlier article, noting, 'The yellow light of gas or candles, so hostile to all *blue* tints, enhances the splendour of red. Ruby becomes more brilliant, *nacarat* [pale red] appears lighter, *cerise* deepens to crimson, and crimson inclines to *capucine*, which itself assumes a more orange-like tone, and orange vies with fire colour' (see Plate 20).[50]

In their creation of vivid images for readers' imaginations, these passages exemplify the mid-nineteenth-century language of colour employed by fashion writers. They also highlight the way in which lighting conditions could significantly affect colours. By considering these changes in colour perception, women could demonstrate their successful negotiation of modern lighting technology, made visible in a pleasing and appropriate appearance.

Conclusion

All of these discussions about colour and textile surfaces illuminate the relationship between women's fashion, science and vision in the mid-nineteenth century. Authors made references to the sciences of vision and light, but also noted effects in terms of magic, underscoring the contemporary awareness that the eye could be deceived. Fashion writers employed a wide vocabulary in their descriptions of colours, alluding to natural phenomena as well as the popular kaleidoscope and encouraging the use of imagination or inner vision. Such descriptions also revealed the contemporary pleasure taken in both colour and language. They discussed colour in terms of remarkable visual displays and also with reference to principles of colour theory drawn from Michel-Eugène Chevreul's publications that recommended or forbade certain colour combinations. Lighting technologies also entered the language of fashionable colour, both in colour names and in advice about how different kinds of light affected different colours. Thus, textile surfaces not only determined the outward appearance and daily experiences of women in the nineteenth century, but also served as examples of the contemporary awareness and appreciation of the scientific and aesthetic qualities and possibilities of the rainbow.

Notes

1 For discussions of seeing and vision from different disciplinary perspectives, see C. T. Christ and J. O. Jordan (eds), *Victorian Literature and the Victorian Visual Imagination* (Berkeley: University of California Press, 1995); J. Crary, *Techniques of the Observer: On Vision and Modernity in the Nineteenth Century* (Cambridge, Massachusetts: MIT Press, 1990); K. Flint, *The Victorians and the Visual Imagination* (Cambridge: Cambridge University Press, 2000); and C. Otter, *The Victorian Eye: A Political History of Light and Vision in Britain, 1800–1910* (Chicago: University of Chicago Press, 2008). Many thanks to Yunah Lee and Lou Taylor for their assistance in preparing this chapter.

2 See, for example, A. Buck, *Victorian Costume and Costume Accessories* (2nd edn) (London: Herbert Jenkins/Carlton, Bedford: Ruth Bean, 1984), p. 19 and C. Goldthorpe, *From Queen to Empress: Victorian Dress, 1837–1877* (New York: Metropolitan Museum of Art/Abrams, 1988), p. 11. Gilles Lipovetsky argues that couturiers such as Charles Frederick Worth played a key role in the growing emphasis upon changes in cut as markers of fashionability in the second half of the nineteenth century (G. Lipovetsky, *The Empire of Fashion: Dressing Modern Democracy* (C. Parker trans.) (Princeton and Oxford: Princeton University Press, 1994), pp. 64–65.

3 Flint, *The Victorians and the Visual Imagination*, pp. 1–2; S. R. Horton, 'Were they having fun yet? Victorian optical gadgetry, modernist selves', in Christ and Jordan, *Victorian Literature and the Victorian Visual Imagination*, p. 3.

4 As Arthur Marwick has observed, the idea that a beautiful appearance reflects inner virtue exists in the Christian tradition, but also stretches back to Plato. See A. Marwick, *Beauty in History: Society, Politics and Personal Appearance c.1500 to the Present* (London: Thames and Hudson, 1988). Valerie Steele and Aileen Ribeiro discuss the connections between morality and appearance, particularly dress. See V. Steele, *Fashion and Eroticism: The Ideals of Feminine Beauty from the Victorian Era to the Jazz Age* (New York: Oxford University Press, 1985) and A. Ribeiro, *Dress and Morality* (new edn) (Oxford: Berg, 2003).

5 [M. Browne,] 'Spinnings in town', *Englishwoman's Domestic Magazine*, NS 2, 15 (1873), 147. Matilda Browne performed both authorial and editorial roles at the *Englishwoman's Domestic Magazine* from 1865 to 1875. Her duties included writing the fashion and advice columns, one of which she called 'Spinnings' and signed 'Silkworm' (M. Beetham, *A Magazine of Her Own? Domesticity and Desire in the Woman's Magazine, 1800–1914* (London: Routledge, 1996), pp. 71, 79).

6 Fictional figures were used to provide useful illustrations, typified by beautiful, selfish Ida Leslie in the short story 'Effie Stanley': 'Fair and lovely, indeed, was the outward form – beautiful the casket; alas, the spirit within so sadly differed!' ('Effie Stanley', *Godey's Lady's Book*, 47 (1853), 22).

7 *Godey's Lady's Book and Magazine*, established in 1830 and published by Louis Antoine Godey (1804–78), enjoyed its heyday from 1837 to 1877, the tenure of its editor Sarah Josepha Hale (1788–1879). The *Englishwoman's Domestic Magazine* was published from 1852 to 1879, edited by Samuel Beeton (1831–77) and

later also by his wife Isabella Beeton (1836–65). Although circulation figures for nineteenth-century periodicals are notoriously difficult to verify, by 1857 the *Englishwoman's Domestic Magazine* professed a circulation of nearly 50,000, having enjoyed 'immediate success after its introduction' ('Preface', *Englishwoman's Domestic Magazine*, 6 (1857), n.p.; Beetham, *A Magazine of Her Own?*, p. 59). By 1860, the *Lady's Book* reached a circulation of 150,000 and was 'the most successful women's magazine to appear in the first three-quarters of the nineteenth century' in the United States (J. W. Tebbel and M. E. Zuckerman, *The Magazine in America, 1741–1990* (New York: Oxford University Press, 1991), p. 33). When quoting directly from the *Lady's Book* in this chapter, American spellings have been retained.

8 J. A. Auerbach, 'What they read: mid-nineteenth century English women's magazines and the emergence of a consumer culture', *Victorian Periodicals Review*, 30:2 (1997), 121. As Auerbach notes, the *Englishwoman's Domestic Magazine* was the first of its kind in England, a publication 'clearly oriented toward the middle-class woman in the home and as consumer' (p. 122). In the absence of a British publication similar to the *Lady's Book* for the period from 1840 to 1848, the *Ladies' Cabinet* was examined.

9 For the history of fashion plates in British and American women's magazines in the mid – to late nineteenth century, see A. Mackrell, *An Illustrated History of Fashion: 500 Years of Fashion Illustration* (London: Batsford, 1997), pp. 132–136, 138–140.

10 Flint, *The Victorians and the Visual Imagination*, p. 2.

11 For an acknowledgment of the continuing difficulties in describing colour, see B. Bloemink, 'A new century of color', in A. Fukai, L. Font, B. Bloemink *et al.*, *Fashion in Colors* (New York: Assouline, 2004), p. 9.

12 [M. Browne,] 'Spinnings in town', *Englishwoman's Domestic Magazine*, NS 2, 13 (1872), 167.

13 'Chitchat upon New York and Philadelphia Fashions for April', *Godey's Lady's Book*, 66 (1863), 407.

14 C. Breward, 'Femininity and consumption: the problem of the late nineteenth-century fashion journal', *Journal of Design History*, 7:2 (1994), 89.

15 M. Schoeser, *Silk* (New Haven: Yale University Press, 2007), pp. 11, 14.

16 'Receipts', *Godey's Lady's Book*, 62 (1861), 461.

17 Schoeser, *Silk*, p. 121.

18 'Flittings', *Englishwoman's Domestic Magazine*, NS 2, 18 (1875), 91.

19 'Paris fashions for the month', *Ladies' Cabinet*, NS 6 (1841), 339; 'Description of the plates', *Ladies' Cabinet*, NS 3 (1840), 442.

20 'Paris fashions for the month', *Ladies' Cabinet*, NS 9 (1843), 338.

21 'Chit-chat of fashions', *Godey's Lady's Book*, 28 (1844), 292. This quotation is taken from Charles Lamb's 1819 essay, 'Valentine's day', in which Lamb writes of a friend creating a valentine, describing the result as 'a work in short of magic. Iris dipt the woof.' The writer for the *Lady's Book* appeared to allude both to the original quotation's reference to a remarkable creation, as well as to the mythological association of Iris with the rainbow and its spectrum of colour. Many thanks to Monica Brewis for identifying this quotation.

22 Crary, *Techniques of the Observer*, p. 20.

23 'Bonnets', *Godey's Lady's Book*, 20 (1840), 220.

24 'Chemistry for youth', *Godey's Lady's Book*, 48 (1854), 566. The article referred to 'nitrate of potass', most likely nitrate of potash.

25 On science and women's periodicals, see S. Shuttleworth, G. Dawson and R. Noakes, 'Women, science and culture: science in the nineteenth-century periodical', *Woman: A Cultural Review*, 12 (2001), 55–70; N. Baym, *American Women of Letters and the Nineteenth-century Sciences: Styles of Affiliation* (New Brunswick, New Jersey: Rutgers University Press, 2002); and A. B. Shteir, 'Green-stocking or blue? Science in three women's magazines, 1800–1850', in L. Henson, G. Cantor, G. Dawson *et al.* (eds), *Culture and Science in the Nineteenth-Century Media* (Aldershot: Ashgate, 2004), pp. 3–13. For two recent treatments of 'popular science' in the nineteenth century that pay particular attention to science as entertainment, see A. Fyfe and B. Lightman (eds), *Science in the Marketplace: Nineteenth-Century Sites and Experiences* (Chicago: University of Chicago Press, 2007) and B. Lightman, *Victorian Popularizers of Science: Designing Nature for New Audiences* (Chicago: University of Chicago Press, 2007).

26 'Moire antique', *Godey's Lady's Book*, 49 (1854), 218.

27 [M. Browne,] 'Spinnings in town', *Englishwoman's Domestic Magazine*, NS 2, 7 (1869), 103.

28 Crary, *Techniques of the Observer*, p. 113; Horton, 'Were they having fun yet?', p. 4. See also I. Armstrong, *Victorian Glassworlds: Glass Culture and the Imagination 1830–1880* (Oxford and New York: Oxford University Press, 2008), p. 255. All these critics discuss Charles Baudelaire's admiration of the kaleidoscope and Crary and Horton note Karl Marx and Friedrich Engels's denunciation of this optical toy.

29 Armstrong, *Victorian Glassworlds*, pp. 255–256.

30 H. Groth, 'Kaleidoscopic vision in late Victorian bohemia: George Sims's social kaleidoscope', in C. Colligan and M. Linley (eds), *Media, Technology, and Literature in the Nineteenth Century: Image, Sound, Touch* (Farnham: Ashgate, 2011), p. 92.

31 J. Gage, *Color and Meaning: Art, Science, and Symbolism* (Berkeley and Los Angeles: University of California Press, 1999), p. 15.

32 P. Ball, *Bright Earth: Art and the Invention of Color* (Chicago: University of Chicago Press, 2001), pp. 175–176.

33 M.-E. Chevreul, *The Principles of Harmony and Contrast of Colors and Their Applications to the Arts* (Faber Birren ed., C. Martel trans.) (West Chester, Pennsylvania: Schiffer, 1987), p. 76. For Chevreul's explicit address to painters and uniform designers, see another English translation: M.-E. Chevreul, *The Laws of Contrast of Colour* (J. Spanton trans.) (London and New York: G. Routledge, 1858), p. 165.

34 Armstrong, *Victorian Glassworlds*, pp. 274–275; Crary, *Techniques of the Observer*, pp. 97–98, 102–107.

35 F. Birren, 'About this book: the French edition', in Chevreul, *Principles of Harmony and Contrast*, pp. 35–36.For an explicit reference to Chevreul in a woman's periodical, see 'Choice of colors in dress; or, how a lady may become good looking', *Godey's Lady's Book*, 50 (1855), 330.

36 Chevreul, *Principles of Harmony and Contrast*, p. 52. For an article that suggests this kind of personal experimentation, see 'Colour, in dress, furniture, and gardening', *Englishwoman's Domestic Magazine*, NS 5 (1862), 59.

37 Armstrong, *Victorian Glassworlds*, p. 267.

38 [M. Browne,] 'The March fashions', *Englishwoman's Domestic Magazine*, NS 2, 12 (1872), 162.

39 S. Milan, 'Refracting the gaselier: understanding Victorian responses to domestic gas lighting', in I. Bryden and J. Floyd (eds), *Domestic Space: Reading the Nineteenth-century Interior* (Manchester: Manchester University Press, 1999), p. 84; J. A. Jakle, *City Lights: Illuminating the American Night* (Baltimore: Johns Hopkins University Press, 2001), pp. 27–28. Milan notes that gas was used to illuminate many public and semi-public spaces, such as streets, shops and theatres, by the 1820s in Britain, but that domestic gas light was not common until the 1870s. Jakle explains that gas light appeared in similar spaces in American cities such as New York and Philadelphia by the 1820s and 1830s.

40 D. de Marly, *The History of Haute Couture, 1850–1950* (London: Batsford, 1980), p. 98. See also E. A. Coleman, *The Opulent Era: Fashions of Worth, Doucet and Pingat* (New York: Thames and Hudson/The Brooklyn Museum, 1989), p. 16.

41 'Levy's in Chestnut Street', *Godey's Lady's Book*, 39 (1849), 297.

42 'London fashions for the month', *Ladies' Cabinet*, NS 2, 7 (1847), 254; [M. Browne,] 'Spinnings in town', *Englishwoman's Domestic Magazine*, NS 2, 4 (1868), 158. For other references to *vert lumière* see, 'Description of the plates', *Ladies' Cabinet*, NS 7 (1842), 136; 'London fashions for the month', *Ladies' Cabinet*, NS 2, 1 (1844), 201; 'Paris fashions for the month', *Ladies' Cabinet*, NS 2, 1 (1844), 272; and 'Paris fashions for the month', *Ladies' Cabinet*, NS 2, 8 (1847), 317. For another reference to gas-green, see [M. Browne,] 'Spinnings in town', *Englishwoman's Domestic Magazine*, NS 2, 5 (1868), 252.

43 'Paris fashions for the month', *Ladies' Cabinet*, NS 2, 8 (1847): 410–411.

44 [M. Browne,] 'The fashions', *Englishwoman's Domestic Magazine*, NS 2, 3 (1867), 480.

45 'Chitchat on fashions for November', *Godey's Lady's Book*, 79 (1869), 463.

46 [M. Browne,] 'The fashions', *Englishwoman's Domestic Magazine*, NS 2, 3 (1867), 89.

47 [M. Browne,] 'The March fashions', *Englishwoman's Domestic Magazine*, NS 2, 12 (1872), 166.

48 Otter, *The Victorian Eye*, pp. 183–184.

49 [M. Browne,] 'The November fashions', *Englishwoman's Domestic Magazine*, NS 2, 15 (1873), 250.

50 [M. Browne,] 'The March fashions', *Englishwoman's Domestic Magazine*, NS 2, 12 (1872), 166.

10 ✧ *Wifedressing*: designing femininity in 1950s American fashion

Rebecca Arnold

NEAT, businesslike, and contained, Anne Fogarty's fall/winter 1955–56 ensemble epitomizes the decade's ideal of efficient femininity (Plate 24). Every detail has been considered. The double-breasted spencer jacket's hem is weighted to ensure that it will sit snugly against the body and the sheath dress is cut to slide across the wearer's curves, echoing her silhouette, but is never too tight or revealing. When buttoned up, the jacket forms a coherent whole with the dress, smooth and grey and ready for the office or a formal meeting. When left open, however, femininity is allowed to blossom, in the form of an ivory silk panel embroidered with grey leaves and flowers. Thus, the outfit embodies the complexity of female identity through its cut and design. Importantly, it also signals hidden, or rather disguised, tensions, through these contrasting surfaces that sit next to the body. The grey wool fabric could be used for a man's light business suit, and is ambiguous in colour and weave, but the yoke's satin sheen and craft-inspired decoration reinforces traditional ideas of femininity. For Fogarty, this play of fabrics, texture and colour was a means to create clothing that would carry women efficiently from office to home, from party to meeting. Both designer and outfit conceal the skill with which gender is negotiated. On the one hand, surface is used to protect the wearer and create an unassailable image of respectable femininity, demure, shapely and modest in its cut and embroidered embellishment; on the other, the menswear fabric and sombre colour speaks of industry, seriousness and discipline. Fogarty was artful in her manipulation not just of her wearers' bodies, but also of the image projected to potential spectators. Her designs embodied 1950s 'surface tensions', in terms of gender, body image and identity, while always seeming guileless in their nostalgia-tinged silhouettes.

During the early years of the Cold War, many involved in the fashion industry displayed a heightened awareness of their role in shaping femininity. Fashion etiquette books had long told women how to dress, groom and conduct themselves, but texts produced by designers during the 1950s foregrounded gender's performativity, whether through garment, text or image, by increasing ideas of empathy and subjectivity, through the author's role as both designer and wearer of her clothing.[1] Photography and styling added to fashion's construction of femininity as something to be created and changed according to occasion, context and intent. Although this can be read as fashion's controlling nature made manifest, magazines such as *Vogue* and designers including Fogarty, Claire McCardell and Jo Copeland continually emphasized women's active role in self-presentation. Trends were never monolithic and a variety of styles were fashionable simultaneously. The 1950s was an era of burgeoning consumerism and plenty, with greater accessibility to goods and wider choice than ever before. The fashion industry thus presented contradictory messages; women were literally 'fashioned' by the advice, clothes and imagery on offer, yet dissonant elements continually undermined the decade's stereotype of women as docile and lacking agency.

By the 1950s, the fashion industry was part of what might be seen as a national project to normalize gender types and reassure the population. Gender was intrinsic to political dictates that concerned domestic life and America's need to assert an easily consumed identity for both home and international markets. Designers, editors and photographers were, consciously or not, part of the construction and dissemination of American femininity. This was a period when government sought to consolidate its strength, in part through a marriage of cultural and political impulses which, at least on the surface, projected a unified set of moral objectives. This essay will explore the significance of surface in designers' approaches to creating ideals of femininity in this context. Anne Fogarty, a New York-based ready-to-wear designer is a key example: she deployed image, object and text to project a feminine ideal, which she also assiduously embodied in her own dress and body image. Her designs focus on surface perfection, and tacitly acknowledge that women have agency in their construction of identity though grooming and clothing. Her work provides a case study of the ambiguous and ambivalent ways fashion can construct gender, and shows how different types of surface can interact to produce a coherent whole.

Laura U. Marks's term 'haptic visuality' is pertinent to the way designers such as Fogarty created designs which invited the onlooker's gaze, while simultaneously combining fabrics and textures to promote the anticipation and the sensation of their touch. This term describes our unconscious responses to the tactile qualities of fashion photography, but also our response to the sight of a dress or fabric. Marks writes of 'touching a film with one's eyes', an impulse that directors can prompt and manipulate. The colours and contrasts in the images and designs discussed in this essay extend this conceit to the fashion page.[2] Anne Friedberg has used a very similar phrase in relation to window displays: 'we touch with our eyes', and fashion merchandising and imagery is designed to tease and entice this inter-sensory complex.[3] During the 1950s, this was heightened within the work of certain designers, who constructed surfaces which, especially when seen in combination with fashionably backcombed and lacquered hair and powdered skin, produced sensual contradictions evocative of visual and haptic dissonance.

Anne Fogarty's advertisements reveal many of the core elements of her work, as well as wider issues within the representation of femininity. An example from 1953 shows a model against a plain studio backing, enlivened only by the addition of a dark chair (Plate 25). As was often the case in the mid-twentieth century, this was joint publicity for the designer and a textile producer. During this period, fabric mattered. Its qualities were extolled with equal, and frequently even more care than those of the designer's work. In this instance, Fogarty's signature curves decoratively around the model's head, but it is the textile that is described in the written panel to her left. While acknowledgement is made that it is Fogarty who 'gives you the perfect-little-lady look', it is the fabric, with its 'happy marriage of dynel, nylon, and finest cotton', that ensures that the garment will remain crease-free, easy to wash and fresh.[4] This readymade dress therefore combines synthetic and organic materials to produce its effect. Yet already there is a tension between ease and rigour, science and nature. The textile manufacturer Bates characterizes this light, plain weave fabric as 'disciplined Challis,' and Fogarty's hourglass design, and the model's awkward pose, indicate that for all the dress's soft folds, the wearer is required to be controlled and dedicated to creating and maintaining this look.[5]

The dress is an archetype of American sportswear.[6] It combines its easy-to-care-for fabric with a classic American style. The shirtwaister evolved from men's work shirts in the late nineteenth century, and evoked an image of efficient, feminine ease. This stereotype was white, middle class and maternal. Fogarty's version is fashionably full-skirted, but not extreme, and falls to below the knee, multiple petticoats exaggerating the fabric's softness. The more dynamic ideal of preceding decades was largely usurped within visual culture by the image of woman as wife and mother. Clothes like this instilled a sense of heritage and tradition, and seemed to reflect a yearning for demure gender ideals that echo those of a century earlier. However, in the postwar period it was difficult to construct such a simple feminine stereotype; even elite fashion magazines recognized that many women worked and had a greater sense of independence than before, even if their eventual goal was marriage and family life. The tensions this caused are present at surface level. While the dress is demurely buttoned and pintucked, it is made from technologically advanced fabric that emphasizes the need for quick cleaning, so the wearer can remain immaculate through a long and busy day. The model's pose is ambiguous. Her eyes are cast down and she refuses the onlooker's gaze. Her hands are clasped and her feet crossed, forcing her to lean awkwardly against the chair. She is perfectly groomed, but her hair is fashionably short and set into curls, which would need only minimal styling. Her skin is smoothed with foundation cream, her eyebrows shaped and her eyelashes mascara-ed and traced with a fine black sweep of liner. Her blusher is discreet. The make-up creates a flawless surface, its 'naturalness' punctured only by the gloss of her red lips and the lacquer-shine of her nails. In Rona Jaffe's 1958 novel *The Best of Everything*, she describes contemporary femininity's complex impulses. Many young women wanted to work, even wanted careers, but were expected to leave their jobs once they married. Set in a New York publisher's offices, Jaffe's book describes the women workers' need to project the correct, groomed image, their neatly feminine ensembles completed by carefully applied cosmetics, the final layer of which was their 'Elizabeth Arden icing'.[7] This coating of light powder ensured that the surface of the make-up remained intact, and that the soft creams and powders cohered to produce a seamless whole.

The image is therefore contradictory, at once reassuring and yet signalling undercurrents of tension. It suggests hidden fears of modernity and change, while signifying both past and present. However, as Mark Jancovich has argued, on closer examination, the decade was driven by inherently contradictory messages; it was 'a period of *both* complacency *and* paranoia, a period remembered for its mindless conformity and for its numerous figures of non-conformity and dissent'.[8] He describes Americans' ambivalent pursuit of conformity, countered by their constant desire simultaneously to cast conformists as 'the other' tainted by associations with totalitarianism and repression. Jancovich sees this conflict as particularly present within the middle class, which adhered to its faith in bureaucracy's 'rational technocracy', while fearing that it would inhibit creativity and lead to mediocrity.[9]

These conflicts were reflected in government policy. This was a period of 'containment' – a much-misused phrase, but one that evokes a cultural, and political, desire to repress, or rather to police, dissent. In global terms this meant American policy sought to vanquish communism both at home and abroad, most graphically demonstrated in McCarthyism; the connection between these aims and consumer culture was made clear in the infamous 'Kitchen Debate' between Nixon and Khrushchev at the American National Exhibition in Moscow in 1959. The policy of containment encouraged constant surveillance, but was tempered by an emphasis on representation. Scott Lucas comments that 'the mission of both the official agencies and their "private" allies was not to contain subversive practices at home ... but to project a positive image of US culture that exalted both "old-fashioned" and "modern" cultural constructions'.[10]

Betty Friedan argued in *The Feminine Mystique* that this led to an almost universally repressive vision of femininity, a view taken up by historians such as Ellen Tyler May, who characterized the period as one of docile domesticity.[11] More recent scholarship has questioned such readings, and called for a more complex understanding of the contradictory impulses embedded within American culture during the Cold War's first fifteen years. However, Friedan's assertion that '[the] mystique of feminine fulfilment became the cherished and self-perpetuating core of contemporary American culture' provides a useful reminder of femininity's centrality to ideas of national identity.[12] Her focus on women, and on the ways big business addressed and shaped their needs, provides a starting point for rethinking fashion in the 1950s. Its designs and imagery seem to reinforce Friedan's ideas but also, crucially, to undermine them, and to suggest that the ideals presented were far from seamless, 'natural' and unquestionable, but actually contained suggestions of their own constructedness and therefore fallibility.

The notion of 'containment' is frequently mapped onto women's bodies. Softer, sportswear styles popular at the time are forgotten, and focus is placed on an archetype drawn from Dior's 1947 New Look. Anne Fogarty's designs sometimes tipped into extreme incarnations of the style, with focus placed on a hard 'body' created by corsetry and shiny dress fabrics that suggested a metallic finish and touch. This was combined with her fetishistic interest in tiny waists – the designer herself boasted repeatedly that hers was only eighteen inches – which added to the sense of a solid, polished, impenetrable surface. This idea of fifties fashion certainly has resonance, even if it does filter out any number of styles that disrupt a clear narrative of the period. Indeed, in Brett Harvey's oral history of the decade, it is the only dress style discussed, and direct links are made between its restrictive forms and political policy. One participant comments, '"Did you ever think about the fact that all the fabrics we wore in the fifties were *stiff*?" My friend Ronnie once asked me. I hadn't, but the minute she said it I thought: faille, shantung, felt, taffeta, pique. Nothing clung, fell, or draped – everything was crisp ... Our cinched waists and aggressively pointed breasts advertised our availability at the same time they warned of our impregnability ... Underneath it all, our flesh, like our volatile sexuality, was "contained" by boned girdles ... in an era when "containment" was a political as a well as a social obsession.'[13] Although this was not the only style fashionable or available, Californian playwear was, for example, very popular, it is perhaps understandable that it was the stiff, uncomfortable clothes and excessive petticoats that dominated memories.

Steven Connor has discussed Michel Serres's idea that the skin is a border between self and world, which extends the body into objects used and connects us to others through touch.[14] In the case of the Anne Fogarty dress shown in Plate 26, the sensual properties of the inner layer of clothing, the restrictive underwear that pinched and heated the skin, are literally and metaphorically extended into the objects it touches. The outfit deploys clashing colours and sensations of soft deep green silk, made shiny and harsh through its satin weave, worn over prickly piles of tomato red nylon net petticoats. The ensemble creates visual, textural and even aural discord between soft and hard, and synthetic and natural. This sensation is then transmitted to onlookers, who would catch glimpses of competing textures, shades and densities of colour as the wearer moved around a party, or got out of a car.

The contradictorily docile, desexualized ideal that these dresses' silhouettes imply has clear parallels with mid-nineteenth-century crinolines, which were frequently depicted in domestic interiors in contemporary fashion plates. This connection between model, fashion and the home was repeated in an Anne Fogarty advertisement from 1954, which showed a model incongruously pouring tea on a plain studio set. Fogarty's design ethos translated into the advice she gave women in her 1959 book *Wife Dressing: The Fine Art of Being a Well-Dressed Wife*. Written in a bright, conversational style, Fogarty advised her readers on what to wear for various occasions, as well as for the various roles that women needed to fulfill, often within the course of a day. 'If I boil down my thinking about clothes into one word,' she wrote, 'that word would be DISCIPLINE – of the mind, the body, and the emotions'.[15] While this rigour is evident in her dresses' carapace-like fit, there is a sense of excess in her work that militates against a simplistic reading. For Fogarty, comfort was a psychological rather than a physical condition.[16] Women might feel appropriately attired in her tight-waisted dresses, but their movements would have necessarily been restricted. As the book's title suggests, in her construction, women were primarily wives, and needed to fashion themselves for lunches and social events designed to further their husbands' careers, even though Fogarty acknowledged that, like the author herself, many had busy working lives too.

A highly successful career woman, Fogarty seems aware of the ambivalence of not just her own position as wife and business woman, but that of her readers. From its opening pages her book makes explicit the constructed nature of femininity and the conscious decisions women must make about how to present themselves in different contexts. This comes across in her constant suggestion that women must recognize the roles they need to inhabit as wives, mothers and, potentially, workers and the ways to manipulate their visual appearance in order to succeed in each of these incarnations. She even described her own tote bag, which is permanently filled with an array of accessories to enable her to change the emphasis and impact of her outfit according to her audience, the time of day and her goals, as 'a kind of conjurer's bag full of magical tricks that help me preserve my appearance and change it at will'.[17]

This performativity is also enacted within fashion pages. For example, a 1949 *Vogue* spread on 'The Apron Era' again owed much to a feeling of excess and exaggeration. The aprons shown in Babs Rawlings's illustrations are notionally practical, yet their frills, ribbons and sheer proportions render them decorative, and even parodic of the 1860s era they reference. They speak more of Marie Antoinette playing at being a shepherdess at the Petit Trianon than they do of genuine domestic preoccupations. Wearers could maintain the surface gloss of pristine domestic organization and traditional identities, while clearly dressed in garments that were merely another layer of feminine role-play. Lary May has described this combination of apparently contradictory impulses as part of 1950s American culture in general:

> In the postwar period, because of freedom, possibilities, and the new sense of liberation, the repressed came back. Those aspects of humanity that had been kept down were accepted and delighted in, for Americans actually desired the returning pleasure they had denied themselves and did not want to relinquish it. So, they simultaneously needed their image of a world in which they have been denied and some way to hold on to what had been repressed, when eventually, it returned.[18]

While May cites film noir and comic books as part of this interplay between hidden, 'repressed' desires and the continued presence of and pleasure in denial, this tension was also contained within some everyday fashions. They spoke of modernity and history, containment and release, through their excess and sensual delight. Illustrations such as Rawlings's can be read as dialectical images: they combine old and new, restraint and decadence, and a fascination with novelty and technology common to both eras, to produce a third image that hovers between the two, incorporating the surface sheen of nostalgia and a knowing glimmer of the present. They contain contemporary fear and anxiety about gender, sexuality and class, and represent the 'return of the repressed' embodied in stiff fabrics and upholstered bodies that parody a feminine natural ideal.

In these examples, the dress becomes the body and the surface becomes the thing itself, as women refashion and remake themselves for each occasion. Both the mid-nineteenth century, which is the designers' reference point, and the mid-twentieth century they actually inhabited, were periods of fraught gender, political and national identity. The first period doubles back, to use Walter Benjamin's ideas, and confronts the present.[19] Pam Cook's discussion of period film costume is also pertinent here. She notes the disruption enacted by its combination of historical and modern elements, comprising an image that is both historicist and current, and thus highlighting its own status as an 'inauthentic' masquerade.[20] This dialectic is present in fabric use too and part of the feeling of excess evident in Fogarty's garments. Her fabric and colour choices demonstrate a preoccupation with surface and contrast, which militates against a simply feminine, demure appearance, and foregrounds clothing's visual and haptic properties. Her constant use of natural and synthetic, past and present within the cut and feminine ideals of her designs is echoed in their dissonant pace of colour and texture. The clothes are constructed as spectacle, foregrounding their artifice to the onlooker and emphasizing the sensual relationship between touch and sight. The prominent textures of the fabrics make clear their tactile qualities, which are there for the onlooker as much as, or perhaps rather than, for the wearer. A striking ensemble from 1957–58 comprises a thickly knitted sheath dress of rich red and yellow checkers with a toning gold-buttoned coat (Plate 27). Designed for formal daywear, this is workwear for upper-middle-class women. The designer's dramatic use of colour undercuts its form-fitting silhouette. Although the wearer might inhabit the male-dominated world of work, which is acknowledged in the outfit's controlled silhouette and meticulous coordination, her difference and 'otherness' is boldly proclaimed in the outfit's hot colours and dense textures.

Another Fogarty dress from 1952–53, this time designed for cocktail hour, made use of the harsh, reflective nature of silver tinsel woven into quilted cloth to produce a glittering full-skirted ensemble.[21] A similar confection is described by Jaffe in *The Best of Everything*; worn by one of the secretaries, who worked in the typing pool on the day of the office's Christmas party, its twinkling fabric and dramatic fit serves as a jolting contradiction to the mundane signifiers of work and industry that surround her glimmering figure:

> There was Brenda at her desk, resplendent in gold lamé, very tight, with a bow right under where she sat. She wore spike-heeled bronze kid pumps and a great many strings of fake beads. She was drinking her morning coffee out of a paper container, leaving a semicircle of lipstick on the rim, and there was a stack of letters at her elbow.[22]

Jaffe describes the outfit in the same terms as the workplace's tinselly decorations, and notes the visual and aural impact of the tactile fabrics worn by other women, which included 'a pale aqua dress of some soft, fuzzy material that looked like angora or rabbit's fur', as well as 'black velvet skirts and white beaded sweaters, or plain taffeta with swishing crinolines'.[23] Such visual and tactile combinations are common to much dress from this period. The garments seem to proclaim their status as costume. Women 'dress up' as mothers, wives, business women, guided by fashion magazines and etiquette books on how to achieve these transformations. They leave traces of themselves, imprints of their make-up on cups and handkerchiefs, echoing clicks of high heels, and the brushing sweep of their skirts as they leave the room. Their presence and absence is bounded by the surfaces that touch their skin and the surfaces and spaces they encounter and pass through.

Importantly, this interest in surface is pronounced even in more monochrome, sportswear ensembles. Claire McCardell's 1957 off-white cotton drill coat, cut in a full, 1920s style, makes use of the heavy weave of the fabric, with its matte tones, as the garment's dominant theme.[24] In contrast, its over-sized buttons were made from walnut shells dipped in metal to produce a high shine gold that glimmered against the plain background. McCardell's interest in colour coordination as a means to mix and match garments from a capsule wardrobe became more dramatic during the 1950s. While she had previously favoured one or two matching colours as a basis for her travel wardrobes and business and leisure ensembles, she now used vibrant clashing hues. Another McCardell outfit from the period brought together an emerald green blouse with a decorative, self-colour weave that added further glowing shine to its synthetic surface, and a cropped scarlet jacket and tight black skirt.[25]

Editorial and advertising photography added a further layer to this dynamic. Its painterly compositions can seem overly contrived and controlled, and like 1950s ready-to-wear it evoked earlier periods, each a mythical 'golden age' of gender differentiation and stability; yet the modern shapes and design forms used undercut a straightforwardly historicist impression. Designers' and photographers' continual focus on surface, texture and colour in the fabrics used, and the ways these are styled and presented, increased readers' sense of dissonance, despite the smooth, seamless finish that magazines strove to construct. Kodachrome imagery made colours glow, and editors revelled in fashion's vivid hues, which were dramatized further with bright, modernist backdrops. Sometimes though, the magazine's picture fractured. The plates used to print the layers of colour slipped slightly to expose the process, rather than the finished product. The image became an abstract pattern, the model's figure visible but hazy, out of focus, like lipstick smeared across the face. This tiny technical flaw revealed the work hidden behind the magazine photograph, and became a metaphor for the discipline women needed to construct their own image. As Fogarty emphasized in all her work, surfaces can be adjusted, manipulated to create a whole, but the elements must cohere. The surface must correspond to the self to be convincing, as Daniel Miller has argued:

> In both philosophy and everyday life we imagine that there is a real or true self which lies deep within us. On the surface is found the clothing which may represent us and may reveal a truth about ourselves, but it may also lie ... There is no true inner self. We are not Emperors represented by clothes, because if we remove the clothes there isn't an inner core. The clothes were not superficial, they actually were what made us what we think we are.[26]

In the 1950s, this meant surfaces that revealed contemporary tensions, which spoke of American femininity in a period of rapid development and flux.

Notes

1 Including, Claire McCardell, *What Shall I Wear? The Who, What, Where, When and How Much of Fashion* (New York: Simon & Schuster, 1956).

2 L. U. Marks, *The Skin of the Film: Intercultural Cinema, Embodiment, and the Senses* (Durham, North Carolina: Duke University Press, 2000), p. xi.

3 A. Friedberg, *Window Shopping: Cinema and the Postmodern* (Berkeley: University of California Press, 1994), p. 66.

4 Anne Fogarty advertisement text, 1953.

5 *Ibid.*

6 'Sportswear, [is] a category of casual but smart separates, dresses and coats, which encompassed resort, campus and career wear for young women ... Sportswear was identified by the wider fashion industry as America's most distinctive form of dress. It comprised three main sub-categories: active sportswear, to be worn, for example, to play tennis or golf; resort wear, which provided clothing to travel in and to wear for leisure and holidays; and town and country wear, most typically tweed suiting.' R. Arnold, *The American Look: Fashion, Sportswear and the Image of Women in 1930s and 1940s New York* (London: I. B. Tauris, 2009), p. 1.

7 R. Jaffe, *The Best of Everything* (1958) (London: Penguin, 2011), p. 57.

8 M. Jancovich, 'Othering conformity in post-war America: intellectuals, the new middle classes and the problem of cultural distinctions', in Nathan Abrams and Julie Hughes (eds), *Containing America: Cultural Production and Consumption in 50s America* (Birmingham: University of Birmingham Press, 2000), p. 12.

9 *Ibid.*, p. 15.

10 S. Lucas, 'Considering cultures: how to make sense of our cold war', in Abrams and Hughes, *Containing America*, p. 191.

11 E. Tyler May, *Homeward Bound: American Families in the Cold War Era* (New York: Basic Books, 1988).

12 B. Friedan, *The Feminine Mystique* (1963) (London: Penquin, 1992), p. 16.

13 B. Harvey, *The Fifties: A Women's Oral History* (New York: Harper Collins, 1993), p. xi.

14 S. Connor, 'Michel Serres' five senses', in David Howes (ed.), *The Empire of the Senses: The Sensual Cultural Reader* (Oxford: Berg, 2005), p. 322.

15 A. Fogarty, *Wife Dressing: The Fine Art of Being a Well-dressed Wife* (1959) (New York: Glitterati, 2008), p. 11.

16 A. Hollander, *Seeing Through Clothes* (Berkeley: University of California Press, 1993), p. 339.

17 Fogarty, *Wife Dressing*, p. 29.

18 L. May, 'Introduction', in L. May (ed.), *Recasting America: Culture and Politics in the Age of the Cold War* (Chicago and London: University of Chicago Press, 1989), p. 31.

19 W. Benjamin, 'Theses on the philosophy of history', in *Illuminations: Essays and Reflections* (New York: Schocken Books, 2007), pp. 253–264.

20 P. Cook, *Fashioning the Nation: Costume and Identity in British Cinema* (London: BFI, 1996), p. 68.

21 C.I.59.16.4, Costume Institute, Metropolitan Museum of Art, New York.

22 Jaffe, *The Best of Everything*, p. 156.

23 *Ibid.*

24 C.I.58.49.8abc, Costume Institute.

25 C.I.58.49.13abc, Costume Institute.

26 D. Miller, *Stuff* (Cambridge: Polity Press, 2010), p. 13.

11 ✧ Long coats, flowing fabrics: fashioning masculinity and desire in film and television

Sarah Gilligan

SCREEN costume functions as a crucial 'storytelling' element of *mise-en-scène*, supporting our understanding of characters' traits, emotions and motivations through a 'dress plot' of changing garments and appearances.[1] Through a combination of fabric, tailoring, surface and movement, costume creates a visual language of its own which possesses the capacity to disrupt and transcend the dominant narrative through melodramatic and spectacular excess.[2] A tension exists between costume serving the dominant demands of the narrative and serving the extravagant possibilities of design.[3] As contemporary film and television become increasingly preoccupied by spectacular excess,[4] garments and accessories draw attention to themselves as an independent source of meaning and pleasure, creating 'spectacular interventions that interfere with the scenes in which they appear and impose themselves onto the characters they adorn'.[5]

This chapter examines the role of screen costume in fashioning masculinity, focusing upon two recent representations of Sherlock Holmes: Guy Ritchie's 2009 film, and Gatiss, Moffat and Vertue's 2010 BBC TV drama series 1.[6] Masculine dress within these texts supports the ongoing reimagining of the character of Holmes, but also creates intrusive spectacular interventions. A complex three-way tension is created within these two texts, between surface adornment, space and the body. In the interaction between garments and the moving body, the former are raised to the level of fetish, offering visual pleasures that classical fetishism cannot fully account for.

Despite the enduring appeal of Arthur Conan Doyle's *Sherlock Holmes* stories with fans, and numerous screen adaptations, it is the cinematic images of Basil Rathbone (1939–46), influenced by Sidney Paget's illustrations of Holmes in *The Strand*, that have come to embody screen, fancy dress and fan representations of the character.[7] Holmes is stereotypically fitted out with a deerstalker hat, caped coat, pipe and magnifying glass, and characterized as 'austere and donnish – lean with sharp eyes, a hawk like nose, a prominent chin and a receding hairline'.[8] Guy Ritchie's film adaptation

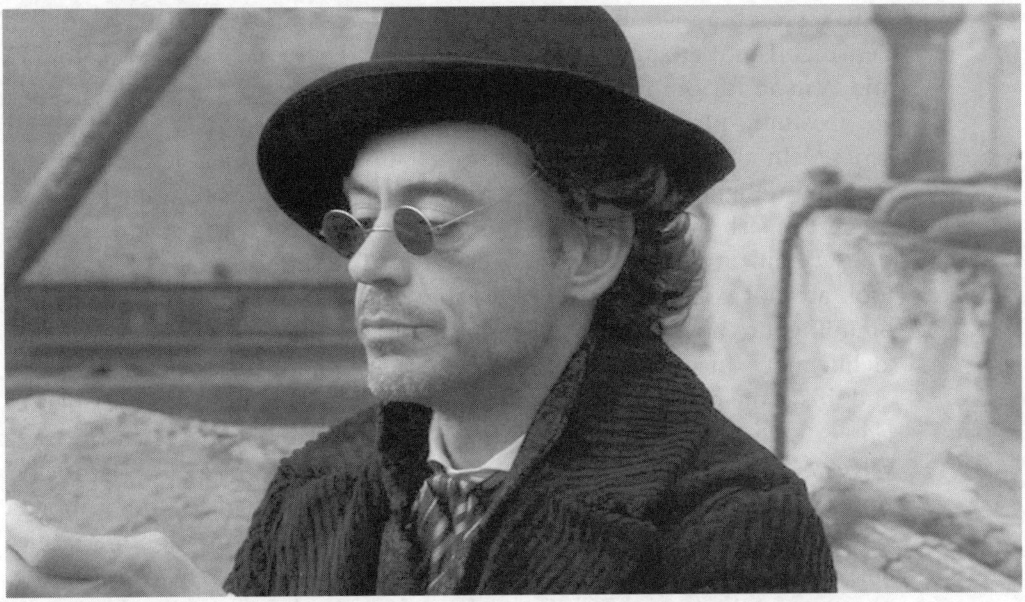

24 Robert Downey Jr as Holmes in *Sherlock Holmes* (2009), directed by Guy Ritchie, produced by Bruce Berman

Sherlock Holmes (2009) and the BBC's TV drama *Sherlock* (2010) both signal a radical reimagining of Sherlock Holmes's 'look' (Figure 24).

Ritchie's film keeps the Victorian setting, but the iconic costume and props are 'replaced – refreshed, some might say – by a borderline iconoclastic, but never anachronistic, bohemian look'.[9] Costume designer Jenny Beavan argues that whilst wholly justifiable, 'for some, it may be a transgression beyond'. The male characters are adorned in an array of quirky bowler hats and fedoras, teamed with tinted glasses and conspicuously showy jackets in tweed, wool, velvet, cord and leather with brash checks and deep jewel colours. The clothes may be 'accurate' in terms of period, but there is a distinctly modern feel about the film through the 'attitude' of the way in which the clothes are worn.[10]

In contrast, Steven Moffat and Mark Gatiss's revisioning of Conan Doyle's stories sheds the Victorian setting in favour of contemporary London. Branded by the BBC as a 'unique', 'thrilling, fast paced update', the three ninety minute episodes for the first series of the programme were aired on BBC One in the summer of 2010. The subsequent DVD release also included a previously unaired pilot episode, offering a subtly sartorially different version of 'A Study in Pink' than that which was aired. The iconic narrative details of Conan Doyle's books remain, with the characters living in Baker Street and 'only interested in the bizarre', but according to Gatiss 'in Benedict [Cumberbatch] and Martin [Freeman], we have the perfect Holmes and Watson for our time' (Figure 25).[11]

Costume plays a key role in creating the 'modern take' for the BBC series.[12] In the promotional discourses surrounding the broadcast series, costume designer Sarah Arthur claims that Sherlock 'would not have any interest in fashion so I went for classic suits with a modern twist: narrow leg trousers and a two-button, slim cut jacket'.[13] The surface details of Sherlock's look are not conspicuously fashionable or spectacular – there is no overt branding, nor any intrusive adornment. Despite this, as Stella Bruzzi argues, 'deliberately unspectacular fashion can still function in a spectacular way' and this is the case with this Sherlock's clothes.[14] The slim cut of his Spencer Hart suit with its deep black surface – which looks like moleskin or heavy velvet – absorbs the light creating a dramatic contrast against Sherlock's/Cumberbatch's pale flesh. His tight-fitting tailored Dolce and Gabbana shirts, with their subtle details of contrasting buttons, shimmering fabrics and open collars, inscribe upon the surface of his body an apparently obsessive attention to detail. Throughout the episodes, the camera repeatedly draws attention to Sherlock's YSL shoes as he runs and offers endless shots of Sherlock putting on and taking off his Belstaff Millford coat (to which I

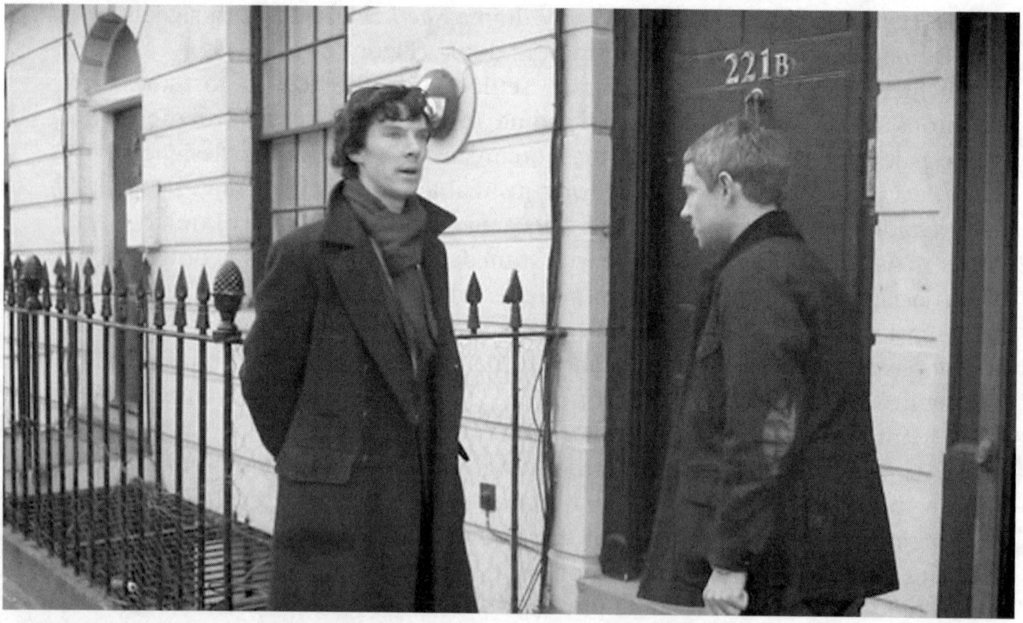

25 *Sherlock* (BBC 2010. 1:1), television series produced by Mark Gatiss, Steven Moffat and Sue Vertue

will return), and his two-tone leather gloves and fine blue wool scarf, both by British designer Paul Smith.[15]

Although it is a minimalist aesthetic, the subtle nuances of fabric, tailoring and surface detail clearly mark Sherlock's difference from the other characters, coding him as both stylish and subtly on-trend. Whilst the character may have no interest in fashion, this reimagining through surface detail is offered up both on screen and in the surrounding promotional discourses as a point of aspiration and desire. Men's style magazines such as *GQ* and newspapers such as the *Guardian* and the *Independent* provide details of how readers can attempt to copy 'Sherlock chic'.[16]

In Ritchie's film, conversely, the pseudo-period cinematic look of Holmes (played by Robert Downey Jr) is characterized by a hotchpotch layering of clothes, often ill-fitting, mismatching, soiled, creased and clashing. This apparent nonchalance both creates a spectacular intervention and is self-consciously emphasized in the film, as Holmes jokingly defends stealing clothes from Watson (played by Jude Law) as part of the 'barter system' between the pair. Holmes's clothes may adorn the surface of his body, but they create a disjointed aesthetic, drawing attention to themselves as seemingly out of place and not in seamless unity with the line of the body, as is the case with Watson's clothes. The way that the clothes are worn upon the body, with a jaunty tilt of a fedora, tinted glasses on the tip of the nose, hands deep in trouser pockets, cord jacket flowing behind and a loosely tied cravat, functions to create a look that is more self-consciously constructed and performed than it may at first appear.[17]

Holmes's studied nonchalance is most explicit in Guy Ritchie's film when he wears his dressing gown (Figure 26). This garment contrasts with the character's impeccable lightweight blue-grey silk dressing gown in the TV series that shimmers in the light and floats gently as Sherlock moves around his apartment. Equally, the cinematic gown lacks the mundane ordinariness and practicality that the garment might be expected to possess – as in the character Arthur Dent's dressing gown in *The Hitchhiker's Guide to the Galaxy*, either in its flowing olive green towelling cinematic representation (2005) or its tied up, bulky burgundy checked fleece incarnation in the BBC TV series (1981). Instead Holmes's gown in the 2009 Ritchie film is characterized by its spectacular threadbare surface. As what Beavan describes as 'the world's oldest dressing gown', the garment, from London costumer's Cosprop, is a 'barely intact paisley' in a rich palette of autumnal oranges and reds.[18] Not only does it contrast with the solid, clean and unmarked deep grey wool coat that Watson wears, but it draws attention to itself both in terms of its patina and its intersection with Holmes's clothed body.

Although patina is most explicitly exhibited on surfaces such as metal and ceramic, it is also evident on the surface of fabrics and materials such

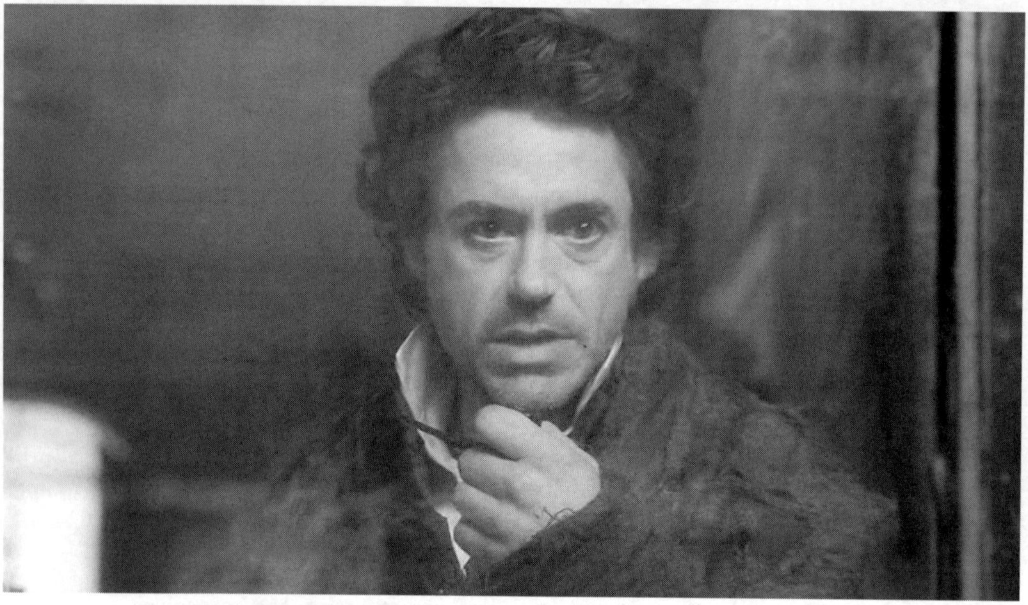

26 Holmes's studied nonchalance in *Sherlock Holmes* (2009), directed by Guy Ritchie, produced by Bruce Berman

as leather. For instance, both Holmes's and Watson's boots and shoes are marked by their contact with the dirty streets of London. Brown leather in particular softens and acquires a distinct mottling as it is worn, an ageing process that invests the garment or accessory with a unique visual and tactile appeal. Yet for delicate fabrics that adorn the body, patina is more often an undesirable marker of wear and tear. Fraying, creasing, fading, tearing, thinning, bobbling, shrinking and stretching combine to create a surface that marks the garment as worn beyond its point of respectability. Whilst in contemporary fashion, denim often has an illusory patina inscribed on its surface, in which wear and tear create a fictional narrative of adventure upon the clothes, the delicate wool cardigan with bobbles and moth holes has outlived its desirability.

Holmes's gown in Ritchie's film is interesting in that it is not constructed from a robust fabric such as towelling, wool or heavy velvet, but features structured tailoring on the lapels and padding on the shoulders, details which make it appear more like a coat than a dressing or smoking gown. It is simultaneously seemingly robust and delicate through its tailoring and threadbare qualities. Its heavy-duty paisley fabric is implicitly invested with status, and its pattern and tailoring imply that it once cost money. The sheer extravagance of the threadbare patina ruffles the surface, creating a complex web of threads and colours that spreads across the formal line and cut of the fabric. Holmes's dressing gown, whilst signifying status through its patina, also constructs the character as more vulnerable than he first appears, functioning as a moment of slippage in his performance of masculinity.

This expression of vulnerability through the details of material surface is also evident in the unaired episode of the BBC's *Sherlock*. In the pilot for 'A Study in Pink', as Sherlock sits in his Baker Street apartment with a cab driver (who proves later to be a serial killer) the camera repeatedly lingers on his shirt collar and neck. As a signifier of masculinity, the shirt collar tends to be represented in terms of its exterior.[19] Watson's collar is tucked in, unobtrusive, coupled with the clean line of a crew neck t-shirt or the protective armour of a thick Arran sweater. Unlike Sherlock, Watson is restrained, confined and contained by his clothes. His collar acts as a residual trace of the structure and conformity of his military past. In contrast, Sherlock's open shirt collar outwardly constructs him as a 'Byronic figure', a seemingly arrogant and alienated remote outsider, a charismatic figure who inspires both fascination and fear.[20] Throughout the (unaired and aired) episodes Sherlock's shirt collar splays out over his jacket, tracing both the shape of his narrow shoulders and drawing attention to his elegant long neck. In the apartment, in the unaired episode, a moment of vulnerability is exposed; the camera lingers not only on the pale flesh of the neck, but also on the interior of the shirt collar – a fragment of cloth which usually remains out of sight. The interior of the collar, a site of intimate intersection between clothes and the body, is marked through sweat, dirt, wear and tear. Sherlock's shirt collar is frayed, so that fragments of pink lining peek through the fabric, highlighting the erosion of surface appearances. The collar's interior is a private surface, an inner space upon which the social life of clothes is inscribed.[21]

In exterior scenes, Sherlock is adorned in a pure wool, Irish tweed, dark blue and grey calf-length coat. With its stiff wide lapels and collar, the heavy wool creates a protective shield that both encases and reveals Sherlock's tall lean body. The collar is worn up at the back, creating a sharp contrast between the solid weave of the fabric and the pale delicate flesh of his face. Even when Sherlock is static, the cut of the coat, especially at the rear, creates the illusion of movement. A half belt strap both creates a focal point of interest, and delineates the conjunction of the smooth curve of Sherlock's shoulders and back and the flow of protruding fabric below.

In the pilot episode, Sherlock's coat is worn in such a way that it wraps and swathes his body. He is shown with his coat done up, and as he walks with his collar firmly up, tracing the line of his face, he swaddles his body with his coat, placing his hands deeply within the coat's pockets to keep it in place. The coat encases his body, its copious fabric at times engulfing

Sherlock's lithe frame. In contrast, in the broadcast episodes the same coat is worn very differently, offering different pleasures. Whilst the collar is still worn up, wrapping his neck, the coat is worn open and allowed to flow freely behind him.

Sherlock's coat, like the ones worn by Neo in The Matrix, the tenth Doctor played by David Tennant in Doctor Who, Captain Jack in Torchwood, V in V for Vendetta (2005) and the title character of Blade (1998) among others, creates an erotic effect as the swathe of fabric floats behind him as he moves.[22] The soft fabric swells, moves and swirls, creating a protective shield, yet one that also reveals a hard, slender body. The self-conscious and spectacular use of flowing swathes of fabric in this way, coupled with slow motion camera work has become a reoccurring stylistic motif in popular film and TV drama since the late 1990s. It is a means of representing male heroes who are 'more sartorially aware and able to combine sharp dressing with fisticuffs and worse'.[23] As I have discussed elsewhere,[24] The Matrix (1999) marked a pivotal shift in representations of the male hero, in which male clothing moved from its role of protecting and concealing to become a central focus for display, for aesthetics over functionality, in this case, a coat. In television drama, shots of well-dressed male heroes with their coats swirling as they are preoccupied in the business of action become frozen in publicity stills, enabling the spectator to revel in the pleasure of the objectified male and the bliss of the textile.[25]

Rather than revealing the bare flesh of the muscular body, such representations often emphasize male physicality via clothes – creating, to borrow from Tim Edwards (in a different context) 'intensely phallocentric' representations in the peek show of 'now you see it, now you don't … [of] his manhood'.[26] Costume is used to create a defensive line between inside and outside, ensuring that the body is read as wholly masculine, rather than possessing the softness attributed to femininity. Yet these flowing coats also generate processes of fetishism and a play of sexual difference. The layers of fabric upon the hard body can be seen to mix both vulval and phallic imagery, constructing Sherlock, like Neo, as an androgynous image made up of both male and female signifiers.

Within media, film and feminist criticism, the appropriation of Freud's concept of fetishism as a theoretical tool can be read as extremely conservative. In 'classical fetishism', sometimes termed 'sexual fetishism', the fetishist is described as pathological or deviant, and active female desire is excluded from consideration.[27] Freud's psychoanalytic interpretation of fetishism centres upon the 'disavowal rather than the pursuit of otherness'.[28] Classical fetish theory would suggest that the coat comes to function as a fetish object, and in turn sexual desire becomes displaced onto a substitute. To Freud, fur and velvet have the capacity to function as fetish objects due to their capacity to 'recall the sight of pubic hair at the moment of undressing,' a moment at which the woman 'could still be regarded as phallic'.[29] Classical fetish theory is most useful when employed to analyse screen costume adorning the body of a woman who is the fetishized object of the active scopophilic male gaze,[30] or when the fashioning of gender is characterized by the use of phallic props, the muscular body or garments which connote robust masculinity such as armour.[31] Classical fetishism does not fully account for a flowing wool coat upon a male body being raised to the level of fetish as the focus of the active female (or same sex) desire. Therefore a broader, potentially more fluid concept of fetishism needs to be applied in exploring the meanings and eroticized pleasures of garments.

Although fetishism is a means of embodying sexuality, power and perception, it is also an increasingly integral part of mainstream fashion. Modern techno-fabrics such as PVC and rubber offer new possibilities for fetishizing the (usually female) body.[32] The wool trench coat is a more complicated case. Though it is offered up as a fetishized object, for the pleasure of either female or male spectators, it is not sufficient as a sexual substitute. The coat as a 'part' does not take the place of the 'whole' person; it is not totally separated as a focus for sexual desire. Perhaps this lack of separation is due to the fact that in the BBC series Sherlock's coat (like Neo's) is not constructed from a conventionally 'fetishistic' material. Its tweed is heavier than the light wool blend that Neo or V wear and is further invested with seemingly magical qualities through bonding with a 'sophisticated ultra-light microporous film' that makes the Belstaff Millford coat waterproof.[33] This techno-fetishist detail is not visually explicit, nor mentioned by any on screen characters, but becomes available through supporting promotional journalistic and fan literature. Such a use of language functions not only to construct the coat as an aspirational, yet virtually unattainable object of desire, but also to displace desire by constructing it as a gadget, or a 'gizmo' like Neo's phone, or James Bond's cars.[34] Despite the alleged 'film' that covers the coat's surface, it lacks the sheen and phallic rigidity of a leather coat, so often associated with cool, confident and rebellious hero characters who exude control, as in the films *Shaft* (1971), the *Wild One* (1953), or *Blade*.[35] Nor does Sherlock's coat appropriate other luxurious, sensuous fetishistic materials such as velvet, silk, fur or lace, which might be deemed to be too frivolous and feminized for contemporary masculinity to wear in the public sphere.

Instead, the sexual charge of the coat is displaced into the power of the fabric to swathe, drape and flow. As Claire Pajaczkowska discusses, cloth is 'a grid, a matrix of intersecting verticals and horizontals, as systematic as graph paper, and yet it is soft, curved and can drape itself into the three dimensional fold'.[36] Because of this flexibility, she writes, 'the extremes of objectification and engulfment feature insistently and repeatedly in our experience and understanding of textiles' in which there 'are surfaces and planes which are known to exist but which are not accessible to view'.[37] Fabric both blurs the boundaries of the body and also frames it through layering. It is at once heavy and mobile. Central to the appeal of cloth is its place between touch and sight, distance and proximity, the seen and unseen, and the known and unknown.[38] Anne Hamlyn similarly argues that 'fabric acts to conceal and cover objects and persons while, at the same time, disclosing them – hinting at their presence'.[39] The wrapping gives the object 'a certain mystery, vitality, and seductiveness. Fabric is malleable. It lends itself to wrapping, draping, and swathing.' It restricts direct access to the naked object, but it also has the ability to suggest, enhance, and draw attention to what it covers over and adorns (Figure 27).[40]

It is difficult to retain critical distance from the fetishistic pleasures of the clothed body in motion; the power of the spectacle proves a repeated distraction. In her introduction to *Undressing Cinema*, Bruzzi writes that when watching the period film *The Piano* (1993) she was moved by the eroticism of the clothes. For her, classical fetish theory provided a useful framework for analysing the displacement of desire onto fabric.[41] Yet in the

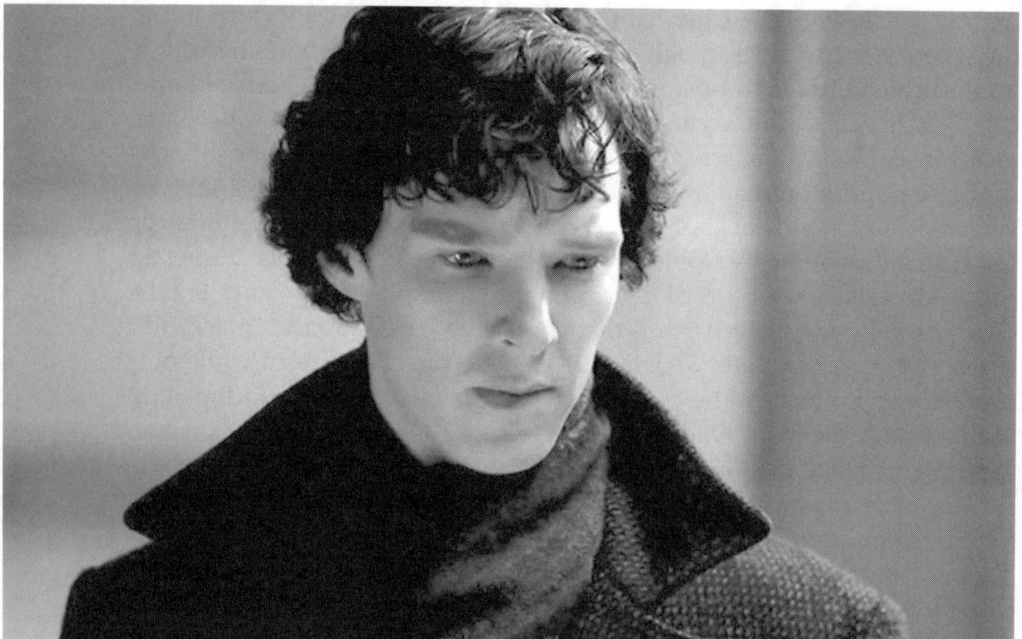

27 *Sherlock* (BBC 2010. 1:2), television series produced by Mark Gatiss, Steven Moffat and Sue Vertue

case of TV drama and film, it is not simply the textile itself which moves and distracts me; it is the relationship of the clothes to the male body. Hamlyn argues that:

> In critical/academic circles, it seems, you either occupy the world of knowledge and insight, commanding language and representation … or you are fated to remain embedded in the seductive world of the sensual, distracted by surfaces and illusions.[42]

Hamlyn attempts to connect these seemingly incompatible fields to Lacan's notion of the Symbolic and Imaginary. For her, fabric occupies both of these fields: '(imaginary) seduction by the surface issue of women's clothes and the (symbolic) distance of the theorising process'.[43] Terming these modes 'touching and naming', she argues that 'much theory has failed to address the relationship between these two positions because doing so might disclose some nasty little secrets'.[44] For her, naming can include a comprehension of touch, but only once truth has been considered from the 'so called feminine point of view' in which women with their seeming 'intimacy with fabric as fetish' may occupy the position of subject for 'critical ends'.[45] In adopting a 'so called feminine point of view' one can argue that fabrics possess their own sensory qualities that enable them to be raised to the level of fetish.

For the fabric to truly act as a fetish object, a sexual substitute, it must exist within the realm of both the visual and the tactile. In the case of screen costuming, the spectator is placed at a distance, unable to touch. Yet point-of-view camera work potentially enables the spectator to revel in acts of touch by proxy, via the implied 'I' of the camera, which stands in for the protagonist. For instance in *The Age of Innocence* (1993) and *The Piano*, the spectator is positioned as a (male) character as he caresses the fabric of a glove, or a tiny hole in a woollen stocking, exploring the boundaries between fabric and female flesh.[46] In the BBC's *Sherlock*, by contrast, although we see the male body dressed, wrapped and swathed in fabric, we frequently do so from the position of an unattributed gaze. We do not survey the garments as Sherlock, or even from the point of view of Watson. As a hero figure, Sherlock is always both subject and object. He is literally untouchable. Perhaps this withholding of tactility is central to constructions of the clothed male as an object of fetishistic desire. Rather than being centred upon the boundaries between fabric and flesh or the revealing of naked flesh, desire is displaced onto the clothes and body in motion. The clothing is a barrier to the concealed body; it is the 'idea', rather than the reality, of male nakedness that heightens arousal. Fetishistic desire thus remains in the imagination, leaving the spectator to endlessly wonder and daydream what the surface pleasures of fabric would be like to touch.

Notes

1 J. Gaines, 'Costume and narrative: how dress tells the woman's story', in J. Gaines and C. Herzog (eds), *Fabrications: Costume and the Female Body* (New York and London: AFI/Routledge, 1990), pp. 180–211.

2 S. Bruzzi, *Undressing Cinema: Clothing and Identity in the Movies* (London and New York: Routledge, 1997); Gaines, 'Costume and narrative'; S. Street, *Costume and Cinema: Dress Codes in Popular Cinema* (London: Wallflower Press, 2001).

3 See Gaines, 'Costume and narrative'.

4 P. Church Gibson, *Fashion and Celebrity Culture* (Oxford: Berg, 2011, pre-publication copy of manuscript supplied to SG by PCG).

5 Bruzzi, *Undressing Cinema*, p. xv.

6 For the purposes of this chapter, I will differentiate between Guy Ritchie's period cinematic and the BBC's televisual representations of the central characters in terms of: Holmes and Watson (Robert Downie Jr and Jude Law) and Sherlock and Watson (Benedict Cumberbatch and Martin Freeman) respectively.

7 For further details see www.imdb.com/name/nm0001651/; see also S. Gilligan, *Fashion and Film: Gender, Costume and Stardom in Contemporary Cinema* (Oxford and New York: Berg, forthcoming).

8 C. Frayling, *Nightmare: The Birth of Horror* (London: BBC books, 1996), p. 200. See also A. Conan Doyle, *The Original Illustrated 'Strand' Sherlock Holmes* (London: Wordsworth editions, 1989 reprint).

9 M. Ryan, 'Sherlock Holmes's costume design on the case of the missing deerstalker hat', *Vanity Fair*, 23 December 2009, www.vanityfair.com/online/oscars/2009/12/sherlock-holmes-costume-designer-jenny-beavan-on-the-case-of-the-disappearing-deerstalker-hat.html, accessed 11 April 2011.

10 Beavan cited *ibid.*

11 BBC 'Sherlock press pack', BBC press office, 12 July 2010, www.bbc.co.uk/pressoffice/pressreleases/stories/2010/07_july/12/sherlock.shtml, accessed 11 April 2011.

12 *Ibid.*

13 Arthur was responsible for the majority of the costumes in broadcast episodes; Ray Holman (costume designer for *Torchwood* and *Doctor Who*) was the costume designer for the unaired pilot. The only garment which was kept from the pilot was the Belstaff Millford coat. See J. Millar, 'How Sherlock became stylish', *GQ*, 11 August 2010, www.gq-magazine.co.uk/style/articles/2010-08/04/gq-style-news-sherlock-holmes-bbc-get-the-look-clothes-watch, accessed 11 April 2011. See also A. Morris, 'Dress like Sherlock and Watson', *GQ*, 25 October 2010, www.gq-magazine.co.uk/style/articles/2010-09/30/gq-style-news-dress-like-sherlock-holmes-and-dr-watson, accessed 11 April 2011.

14 Bruzzi, *Undressing Cinema*, p. 25.

15 Morris, 'Dress like Sherlock'; Millar, 'How Sherlock became stylish'.

16 Morris, 'Dress like Sherlock'; Millar, 'How Sherlock became stylish'; P. Bignell and R. Shield, 'Sherlock chic! Fashion? It's elementary, thanks to BBC', *Independent*, 8 August 2010, www.independent.co.uk/life-style/fashion/news/sherlock-chic-fashion-its-elementary-thanks-to-bbc-2046500.html, accessed 11 April 2011; A. Petridis, 'No chic, Sherlock', *Guardian*, 4 September 2010, www.guardian.co.uk/lifeandstyle/2010/

sep/04/sherlock-fashion-mens-coats, accessed 11 April 2011; see also Gilligan, *Fashion and Film*.

17 See also C. Laverty, 'Sherlock Holmes costume guide part 1: frock coats and bustles', *Clothes on Film*, 17 May 2010, http://clothesonfilm.com/sherlock-holmes-costume-guide-part-1-frock-coat-bustles/10814/, accessed 25 April 2011.

18 Beavan, quoted *ibid.*

19 See R. Dyer, *Heavenly Bodies: Film Stars and Society* (London: British Film Institute, 1986).

20 See A. Stein, *The Byronic Hero in Film, Fiction and Television* (Carbondale: Southern Illinois University Press, 2009); A. Spicer, *Typical Men: The Representation of Masculinity in Popular British Cinema* (London and New York: I. B. Tauris, 2003).

21 R. Healy, 'Fractured fairy tales', 'Fashion in Fiction' conference, University of Technology, Sydney, 26–27 May 2007; L. Mason, 'The social life of clothes', 'Fashion in Fiction' conference, UTS, 26–27 May 2007.

22 S. Gilligan, 'Becoming Neo: costume and transforming masculinity in the *Matrix* films', in P. McNeil, V. Karaminas and C. Cole (eds), *Fashion in Fiction: Text and Clothing in Literature, Film and Television* (Oxford and New York: Berg, 2009), pp. 149–160; S. Gilligan, 'Fashioning masculinity and desire', in A. Ireland (ed.), *Illuminating Torchwood: Essays on Narrative, Character and Sexuality in the BBC Series* (Jefferson, Missouri and London: McFarland, 2010), pp. 153–164; Gilligan, *Fashion and Film*.

23 P. Church Gibson, 'Queer looks, male gazes, taut torsos and designer labels: contemporary cinema, consumption and masculinity', in B. Babington, A. Davies and P. Powrie (eds), *The Trouble with Men: Masculinity and Contemporary Cinema* (London: Wallflower Books, 2004), p. 180.

24 Gilligan, 'Becoming Neo'.

25 *Ibid.*; Gilligan, 'Fashioning masculinity and desire'.

26 T. Edwards, *Men in the Mirror: Men's Fashion, Masculinity and Consumer Society* (London: Continuum, 1997), p. 42. Within *Sherlock Holmes* the spectator does briefly see the bare flesh of Holmes's body as he fights in a sequence reminiscent of *Fight Club* (1999) (see Gilligan, *Fashion and Film* for further discussion).

27 A. Fernbach, *Fantasies of Fetishism: From Decadence to the Posthuman* (New Jersey: Rutgers University Press, 2002); L. Gamman and M. Makinen, *Female Fetishism: A New Look* (London: Lawrence and Wishart, 1994).

28 Fernbach, *Fantasies of Fetishism*, p. 4.

29 Freud, quoted *ibid.*, p. 7.

30 Bruzzi, *Undressing Cinema*.

31 Y. Tasker, *Spectacular Bodies: Gender, Genre and Action Cinema* (London and New York: Routledge, 1993); R. Dyer, 'Don't look now: the male pin-up' (1982), in *Screen* (ed.), *The Sexual Subject: A Screen Reader in Sexuality* (London and New York: Routledge, 1992), pp. 265–276; S. Neale, 'Masculinity as spectacle' (1983), in *Screen, The Sexual Subject: A Screen*, pp. 277–290.

32 Fernbach, *Fantasies of Fetishism*; V. Steele, *Fetish: Fashion, Sex and Power* (New York and Oxford: Oxford University Press, 1996).

33 See C. Campbell, 'Belstaff keep it Sherlock real', *London Daily News*, 3 November 2010, www.thelondondailynews.com/belstaff-keep-sherlock-real-p-4715.html,

accessed 25 April 2011.

34 R. Banham, 'The great gizmo', in R. Banham and M. Banham (eds), *A Critic Writes: Essays by Reyner Banham* (Berkeley, Los Angeles and London: University of California Press, 1996), pp. 109–118; S. Gilligan, 'Get me an exit: mobile phones and transforming masculinity in the *Matrix* films', in R. Chueng (ed.), *Cinema, Identities and Beyond* (Newcastle-upon-Tyne: Cambridge Scholars Publishing, 2009), pp. 54–70.

35 J. Harvey, *Men in Black* (Chicago: University of Chicago Press, 1995); Bruzzi, *Undressing Cinema;* S. Gray, 'Shaft's leathers: the masculinity of John Shaft in Gordon Parks' 1971 film, *Shaft*', 'Fashion in Fiction' conference, UTS, 26–27 May 2007.

36 C. Pajaczkowska, 'On stuff and nonsense: the complexity of cloth', *Textile: The Journal of Cloth and Culture,* 3:3 (2005), 233.

37 *Ibid.*, 224.

38 *Ibid.*, 224.

39 A. Hamlyn, 'Freud, fabric, fetish', in *Textile: The Journal of Cloth and Culture,* 1:1 (2003), 11.

40 *Ibid.*, 11.

41 Bruzzi, *Undressing Cinema,* p. xiii.

42 Hamlyn, 'Freud, fabric, fetish', 11.

43 *Ibid.*, 11.

44 *Ibid.*, 11.

45 *Ibid.*, 11.

46 Bruzzi, *Undressing Cinema;* Hamlyn, 'Freud, fabric, fetish'.

12 ✧ Surface as material, material into surface: dialectic in Carol Christian Poell

Ulrich Lehmann

Surfaces are dialectic. Structurally they appear as boundaries between the materiality of an object and the empty space that surrounds it; they demarcate one material substance from another non-material one. In circumscribing the dimension of an entity they provide the limit of its expansion. Surfaces determine perception by constituting an outer layer that can be seen as an integral part of the object's volume and plasticity. But when distinguished conceptually from that object, they become its antithesis: an insubstantial *superficies* (also the etymological base from which the word 'surface' has emerged). The surface as thesis, as exterior limit, has as its antithesis the integrity of texture, 'finish' and form.

As an outer layer, the surface integrates with the object through material congruence or, conversely, through the 'falsehood' of masking the plastic values of one material underneath another. Historically this 'falsehood' has been understood as negative when surfaces signal values that are not realized in the substance beneath them.[1] A bronze patina on an iron statue or the silver on Sheffield plate are examples of surfaces as material pretence, suspect to deceive. In modernity such pretence has been explored as emerging from a distinct socio-economic and cultural value system, which is built on insubstantial, rapid production and constantly renewed fashions. Subversive or ironic reversals of surface deception become infused with creative potential, for instance in Carlo Bugatti's furniture with its parchment and pewter details. Mid-twentieth-century veneered furniture in which chipboard is adorned with layers of acrylic lacquer, or the intricate 'curtain' walls made from expensive glass that are hung before architectural grids of raw iron beams, are instances in which the surface defines the integrity of the object, substantiates in material, economic and aesthetic terms. As antithesis, the surface possesses distinct material qualities – varying degrees of thickness and density which allow it to be defined and explored as an object in its own right.

At this point a brief etymology of the word 'surface' can be useful. Its earliest occurrence in Romanic languages has been recorded in 1378, when the poet Jean Le Fèvre used it as synonymous with 'partie extérieure'. The lexicographer Antoine Furetière defined the mathematical surface in his *Dictionnaire Universel* of 1690 as 'étendu plane [extended plane]'.[2] English usage ranges from similar planar models – 'magnitude or continuous extent having only two dimensions (length and breadth, without thickness)' – to emphatically material terms: 'the outermost part of a material body, considered with respect to its form, texture, or extent; the uppermost layer; [especially] in art or manufacture, an exterior of a particular form or "finish"'. There is also the sense of surface, in terms of perception: the 'part or aspect of anything which presents itself to a slight or casual mental view, or which is perceived without examination: outward appearance'.[3] The move from a material to a metaphorical or in this case, significantly, a phenomenological meaning is due to habitual linguistic and cultural development that progresses from denotation to connotation. For the present discussion this shift in meaning reflects a movement from essentialist, idealist tenets toward dialectical perception. What concerns us is not the simple extension of the material constituent 'surface' toward 'superficiality' as a symbolic value judgement, but rather the perception of the 'finish' as antithetical to the material substance that nonetheless is required to complete it. Instead of assuming that a *superficies* reflects (at times literally), conceals or detracts from the object beneath it, the surface is here posited as a construction that determines what is interior to it.

It is important not to set exteriority and interiority within a hierarchical value system, for instance by assuming the idealist position of subjectivity as superior to extraneous material. What is 'superficial' envelops and demarcates the interior space, is put above it linguistically ('super'/'supra'). It is a hull or front that determines the object as material thing, with corresponding meaning and perceptions. But, conversely, the surface is not superior in significance to the object or body it clothes. Its finishing of the object does not constitute an operation that promotes visual or haptic appearance over interior construction or constitution.

Extending plane (dialectic)

The oscillation between surface value and material object is also a significant analogy for the dialectical relationship between consumption and production. Just as a produced object is transformed into a commodity by being embedded into a socio-economic value system, the surface 'finishes' the object by integrating it into a phenomenological relationship with the beholder. The tactility of the surface primes the viewer or user for a series of actions that complete the essential meaning of the object. Such an operation is dialectic, too. As the consumption of the surface determines the construction of the object, the making of the object includes within its production from the start the 'finish' or surface – even (or especially) when object and surface are stood in material contrast. Karl Marx stated the dialectic of 'consumption' and 'production' as follows:

> Consumption produces production in a double way, 1. because a product becomes a real product only by being consumed. For example, a garment becomes a real garment only in the act of being worn ... Only by disintegrating the product does consumption give the product the finishing stroke;[4] for the product is production not as objectified activity, but rather only as object for the active subject; 2. because consumption creates the need for *new* production, that is it creates the ideal, internally impelling cause for production, which is its presupposition. Consumption creates the motive for the production; it also creates the object which is active in production as determinant aim.[5]

For purposes of our analogy this passage implies that the 'surface' (consumption) presents the 'impelling cause' for the material (production). Without anticipating the surface the object cannot constitute itself as material. If, as Marx proposes, the garment only becomes a garment when worn, then implicitly, esoteric, individualized dress only becomes 'fashion' when presented in a distinct socio-economic context (social competition, gender performance, etc.). It is obvious that garments only constitute themselves as clothes when they actually dress the human body; the material design is pointing from the beginning at its use as sartorial surface. From this example we can extend easily the logic that surfaces determine the perception of objects in both phenomenological and material terms. But Marx also makes clear that the product disintegrates through the act of consumption, in the same way that fashion 'dies' when a once-new stylistic proposal is consumed by a larger portion of society and has to be reborn through a new style, a subsequent act of consumption.[6] Similarly, the surface disintegrates the substance of the object through its elevation of the boundary toward negative space over the positivist constitution of volume or matter. This disintegration has an ontological quality, which can be traced in the habitually negative connotation of 'superficial' as moving above or beyond the face or form into a domain of the void – the loss of meaning and denotation, the dissolution of structural substance into a semantic or visual front.

The post-structuralist/post-Marxist discourse put forward by Gilles Deleuze and Félix Guattari in the two volumes of their *Capitalisme et schizophrénie* (published 1972 and 1980) develops the notion of the 'surface' from the root (not yet the rhizome) of the fundamental dialectic outlined in Marx's *Grundrisse*. Here, the 'disintegrating [of] the product' through consumption is extended into the notion of 'visagéité' (faciality) that ushers in a process of deterritorialization. This is built upon the simultaneous rejection of 'significance' and 'subjectivation', meaning and subjectification, as the twin pillars of idealist construction. Deleuze and Guattari do not move from their refutation of idealism toward avowed materialism, but rather into the realm of deconstructed signification that eschews all subjective and phenomenological reference in order to bare the operating mechanism (the 'facial machine') of reification within capitalist economies and its psychosomatic impact.[7]

In their work the notion of 'surface' can be dissected into two parts: 'sur' (beyond) and 'face'. This juxtaposition signifies an extension from surface structures into corporeal, sexual and socio-political complexes. For Deleuze and Guattari 'the face is part of a surface-holes, holey surface, system [*système surface-trous, surface trouée*]'. Eyes and mouth punctuate the unanimity of the 'white wall' of the facial surface and structurally inscribe themselves on a blank page.[8]

> This system should under no circumstances be confused with the volume-cavity system proper to the (proprioceptive) body. The head is included in the body, but the face is not. The face is a surface: facial traits, lines, wrinkles; long face, square face, triangular face; the face is a map, even when it is applied to and wraps a volume, even when it surrounds and borders cavities that are now no more than holes. The head, even the human head, is not necessarily a face. The face is produced only when the head ceases to be a part of the body, when it ceases to be coded by the body, when it ceases to have a multidimensional, polyvocal corporal code – when the body, head included, has been decoded and has to be *overcoded* [*surcodé*] by something we shall call the Face [*Visage*].[9]

The Marxist dialectic of production as distinct from, but entirely dependent on, consumption here finds a late reprise through the notion of the 'face' as structurally distinct from but codified in the body. The term *visage* is the expression of a coded production of identity that is 'sur'-human. This permits a reading of the dialectic of surface and material wherein the former appears as the coded production of the latter under certain socio-economic conditions of labour.

Despite Deleuze and Guattari's evocation of the human face I want to resist the temptation to slide from the dialectic between object and surface to an 'original' dialectic between the body and its skin during human physical and mental development. I do not intend to recount the investigations by Abraham or Anzieu, who put forward a psychoanalytical interpretation of the corporeal surface in the 1970s;[10] and in terms of methodology, I do not want to move the present argument into the realm of psychoanalysis but remain largely within a material framework. I want to refrain from the subjectification that psychoanalytical theory and praxis necessarily imposes, and explore instead material structures as primers for social ones: what has been termed 'reification' or 'objectification'. I develop this theme from the historical dualism of subject<->object (man<->nature) into a dialectic of surface/material<->object, which flickers constantly as signifier/signified, a conceptual oscillation that drives the reification process of the modern subject. The relationship of the subject to the object shifts constantly with regard to both material and surface. Physical and psychological reactions (even psycho-pathological ones) are indeed caused by the shift between the two and cannot be observed only phenomenologically. Guattari and Deleuze, in their twin rejection of signification and subjectification were adamant that the body has to be taken out of the concept of faciality (their variant of reification):

> The semiotic of the signifier and the subjective never operates through bodies. It is absurd to claim to relate the signifier to the body. At any rate it can be related only to what has already been entirely facialized. The difference between our uniforms and clothes and primitive paintings and garb is that the former effect a facialization of the body, with buttons for black hole against the white wall of the material [*étoffe*].[11]

Here the de-subjectification is allocated to clichéd uniformity (literally in the wearing of uniforms rather than civilian garments). But subjectivity is equally absent in fashion's dress codes – although it might be upheld through certain, limited esoteric choices. The garment, through the reifying process of its commodification *as fashion*, decrees uniformity just as much as a concretely coded uniform, but in a much subtler and perhaps a more cynical form.

Against subjectification

The surface of fashion, in fabrics [*étoffes*], textiles, accessories, make-up and hairstyles, is a repetitive subjectification of dress codes through stylistic manoeuvres that give form to established garment types – as there exists only a limited range of viable covers for the human body – and create new objects for the consumer. The consumption of surfaces is also the production of clothes as fashion-objects. Surfaces therefore fulfil Marx's first postulate in making the garment wearable: in establishing as a tactile boundary the relationship between body and sartorial shell. But, in line with Deleuze and Guattari's ontological renunciation of the body in reification, the wearer disintegrates beneath the fashion surface. The body becomes a system of signification and the materiality of limbs, torso and sexual organs is covered up by the 'faciality' of a textile front. This front thus moves from textile to text. It can be read as signifier for the particularized subject-object relation that is performed within the modern code of confrontation between idealist traditions and material reality.

But the desire for subjectivity remains constant despite its material impossibility. Therefore the dialectic of surface and material object remains valuable for the categorization of the subjectifying impulse – most notably in fashion.

Due to both the physiology of the human body and the anthropological necessity for protection and modesty, clothing can adopt only a finite set of material forms. Fashion therefore needs to alter its *superficies* in order to sustain its consumption. This occurs first on the surface of clothes and accessories, through delineating and decorating. By this I mean the act of circumscribing the body through outlines, i.e. the silhouette itself, but also articulating aspects like piping or hemlines, which indicate the 'beginning' or 'end' of a clothing design, structuring devices like seams and darts, and the visual character of the surface itself, which may create the illusion of continuity or interruption.

In this context the word 'pattern' carries a double meaning. It pertains in habitual parlance to the embellishment of a textile outer layer through repeated motifs that can travel across the traces of manufacture by connecting fabrics on either side of seams or darts, thereby unifying a piece of clothing.[12] Simultaneously, the 'pattern' is the principal component of garment construction – the basic shapes that are cut in order to be tailored. Here, one finds again the double bind of textiles, as signifying surfaces (carrying of symbolic or aesthetic meaning) and as material building blocks for the circumscription of human physiology.

The 'tension' between perceiving surfaces either as subjectified phenomenological, aesthetic or symbolic markers or as materially innovative objects expresses itself in fashion also through industrial working processes. The largest part of global garment production – especially in the mid-price and economy ranges of clothing – operates on the principle that surfaces have to be decreed before fashion can be manufactured. This functions through trend- and colour-forecasting and the 'futurizing' of fabrics. Designers will congregate at annual trade fairs and consult trend books in order to decide at least a couple of seasons in advance on the basic materials and colour schemes for their collections.[13] Subsequently, shapes and forms, cuts and fastenings, etc. are decided on the basis of a nominal 'theme' or 'concept'. Identical or near identical designs by different manufacturers can be demarcated by the imprinting of logos onto the surface of the clothes. These can appear as actual registered trademarks (for instance interlinked 'C's for a traditional French label and the 'LV' of a luggage manufacturer) or as decorations that repeat and vary across a range, for instance the swirling Pucci-patterns of the 1960s or Burberry's traditional glen plaid. These signifiers both establish the visual character and, more significantly, brand these products for their consumption (and eventual disintegration). They can be used to integrate a whole range of diverse designs – even if they appear atypical to the 'tradition' of the fashion company – into a unified corporate image. Here again, the surface appears to negate the materiality of the clothes through its transfer into coded signification, while also constituting its concrete material value. Identical t-shirts (often produced in the same outsourced factory) can be 'produced' by a succession of diverse fashion labels through the differential logorization of the surface.

In order to employ a material approach that complements the structural interpretation above and, at the same time, points toward a concrete application of the surface/object dialectic to existing fashion work, I would like to discuss in the following section some examples of work by Carol Christian Poell. Born in Linz, Austria in 1966, Poell studied fashion design in Graz, Vienna and Milan, where he founded, together with Sergio Simone, his company C.C.P. srl. Since 1994 his studio has created clothing without being constrained by seasonal rhythms or gender boundaries, in cooperation with small artisanal producers in Japan and the north of Italy. Poell's family background – his grandfather and father were tanners – and his traditional apprenticeship in tailoring offer a particular approach to the making of clothes and accessories. I will employ the notion of the 'surface' and its 'tension' between material object and signifier (boundary to the void, ground for the inscription of texts) to analyse Poell's 'pieces', particularly with regard to his treating of leather.[14]

Soiled surface

My first case study is a pair of leather trousers, dyed in bull's blood and fitted front to back (Figure 28). They were designed in 1999 as part of Poell's series 'Fe-Male'. The surface articulation is here painterly as much as it is material. The process of dyeing the trousers in coagulated blood is documented in the individual traces that the application has left on the leather – individual and subjective gestures that are akin to the works of *art informel* or *tachisme* where the properties of the paint and the direction of the hand (pouring, smearing, brushing) produce improvised and seemingly random layers and articulated textures of paint. At the same time the bull's blood not only penetrates the grain of the leather, imbuing it with a particular colouration and metallic scent, but builds into a surface texture that works simultaneously with and against the leather – one layer atop another.[15] The first layer here is the sartorial surface; the fe/male wearer sporting a pair of trousers in a material

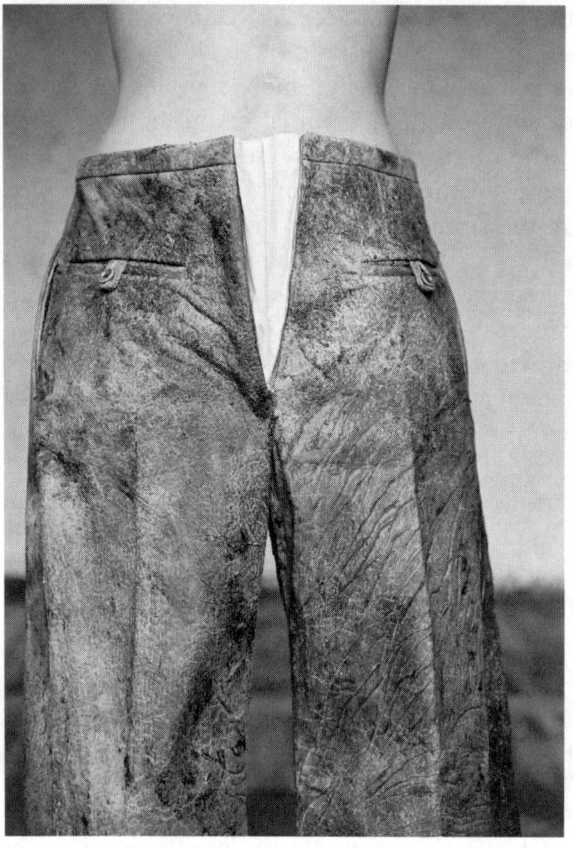

28 Carol Christian Poell, bull's blood-dyed leather trousers fitted front to back, 1999

that is habitually decreed by fashion as eminently sexualized: an animalistic skin in the 'raw' which, despite the treatment of the material (tanning, dyeing, etc.), is opposed to the human, refined skin that is subject to beauty treatment. These two skins are articulated as surfaces through their innate physiological relation – the first and original boundary between an animal and its surroundings – but the upper layer (the leather) is posited also as the dead artifice that now provides warmth and protection to the living surface of the human epidermis. In coating the leather with blood drained from a slaughtered bull, Poell simultaneously animates and reifies its surface; it is akin to the living skin of the subject who wears the garment, but constitutes at the same time an objectified cover that approximates an artistic gesture.

Not merely because of his Austrian birth, Poell's use of blood as surface coating must refer the 'Orgien Mysterien Theater', which was staged by the Viennese artist Hermann Nitsch from 1962 onward.[16] These performances of partly clothed participants, who were covered with animal blood and intestines, constructed contemporary artifices of primary – albeit often fictional – rituals of social initiation and survival. Nitsch's animistic iconography and his use of materials and modes of presentation that are largely extraneous to art, approximate the archaic and universal gestures of a religious mass or violent protest. Applied to the commodified world of fashion, this actionism is translated into gestural strokes of blood on the leather trousers, which similarly affect connotations of ritual.

The pattern cut for the trousers is reversed along a side axis so that they are fitted to the front of the wearer while open in the back (whose position is further articulated by the traditional two buttoned 'back'-pockets). This reversal draws attention to the habitual gender distinction within the cut of conventional trousers and articulates the sex by transferring the mechanism for potential exposure from one erogenous zone to another – from genital area to anus. The surface of this inversely assembled pattern, in dressing the nether regions of the fe/male body, reveals itself again as dialectic in articulating physiological mechanisms and functions of the body while at the same time negating this skin/surface through the expressive materiality.[17]

Surface as object

The second case study from C.C.P. concerns a leather jacket designed in late 2007 (Figure 29). In this garment leather is mainly treated as an 'essentialist' surface. Its origin as a hide that stretches across animal bodies renders it, structurally speaking, a cover: its primary function is to provide protection (warmth, camouflage, etc.) and only on a secondary level does it perform connotative functions of species identification or gender specificity. But the expansive and complex process of treating hide for fashion consumption through depilating, tanning and dyeing, and the articulated and hard texture lends leather garments or accessories distinct object value. For Poell, too, leather is therefore surface as well as material. Its character as a cover, which initially appears to negate its object status, is conversely emphasized as a material constituent through the reversal of the traditional working process.

The kangaroo leather for this short jacket is object-tanned and – dyed. This is a mainstay within Poell's technique of processing leather and it originates partly in the familial connection to his ancestor's tannery in Austria. Object-tanning and -dyeing act as departure points for an ingenious array of processes in which Poell experiments with the properties of the hide. Significantly, these properties are not defined by their final appearance within the garment – as surface texture or colour, for instance – but are analysed and creatively articulated during the preparation process. It is important to underline that this is not the usual approach to treating a hide, which is increasingly worked in a similar fashion to fabrics like wool or cotton (dyeing, printing, etc.). Poell instead takes the original, basic

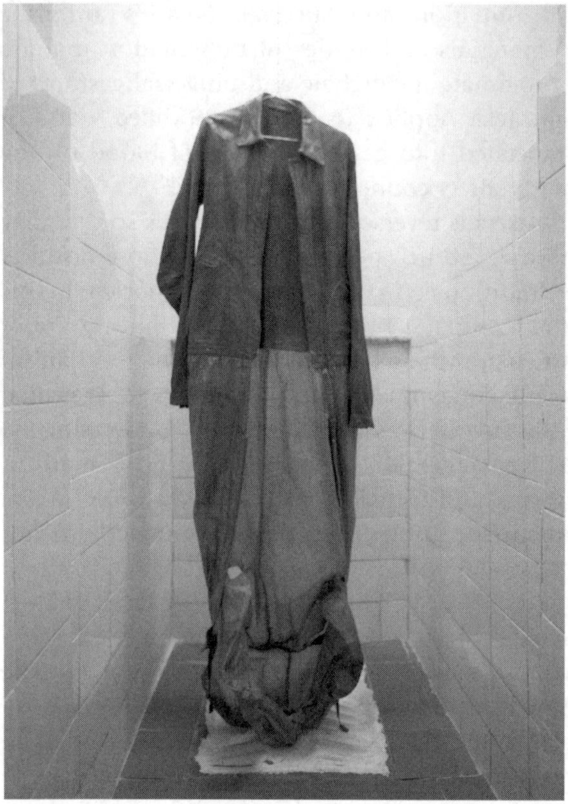

29 Carol Christian Poell, jacket of kangaroo leather, 2007

properties of the leather and uses them to guide a particular construction for the pattern pieces, which in turn constitutes – extraneous to any particular style or look – the character of the assembled piece.

Initially, the hide is merely depilated and rendered pH-neutral. The pieces that are cut from it have to be measured with a margin of between seven and ten per cent to allow for the shrinkage that will occur in the tanning and dyeing processes to follow. During cutting and stitching the raw hide is kept constantly moist by resting it on water-soaked mattresses next to the sewing machines; overnight it is refrigerated. Were the piece to be left at this stage it would indeed resemble a finished leather garment with all seams and fastenings in place but it would also 'fossilize' within hours, becoming stiff like the skin of a mummified corpse. Only after the completion of the garment is it 'finished' through subjection to the tanning and dyeing that normally occurs at the very start of the process. The surface treatment thus affects the whole garment and allows it to be comprehended structurally *as object* generatively and not as a projected assembly of pattern pieces and details.

The kangaroo hide used for the jacket is thin and malleable but also rather small in its individual surface area (compared to cow or horse), so the overall pattern has to be adapted carefully. At the end, the jacket is fully 'lined' with a nylon 'double' that can be pulled out and folded up and over the leather from the bottom back hem of the garment to provide it with yet another, waterproof surface.

Healing skin/surface

The conceptual rigour in Poell pieces is evident in the length of time it takes to research and complete particular objects and garments. This extension, reaching across seasonal presentation and commercial deadlines, occurs equally in the narratives for the series of pieces, which seem to reflect on previous topics within the designer's output or point forward to a developing design process. Thus works for the first part of the *Trilogy of Monotypologies,* dating back to the autumn of 1999, which featured the notion of bleeding or of being bled, were continued in 2000 with a series that materialized the stemming of blood flows and the healing of wounds.

The *bodyplasts,* which resemble in material quality, surface texture and colouration enlarged band-aids that adhere to the skin, suggest the need to heal wounds on the wearer's body (Figure 30). Whether these wounds are metaphorical, as indicators of discrimination and alienation of the body or real marks and traces of daily conflict on the physiological surface, the effect of the bodyplasts is not comforting. They are too large, too deliberate in their covering up to be assisting a healing process. They seem to proclaim a festering and permanent injury that cannot be contained let alone cured.

The blood that has flowed over trousers and skirts in Poell's previous pieces cannot be stemmed by such bodyplasts. One suspects that, once these textures are removed and the skin is painfully rid of such an adhesive surface, its epidermis would open up again, old injuries break out, and blood would continue to drain from the body. Were the injuries metaphorical they might well constitute socio-cultural stigmata, to which the bodyplasts draw further attention in distinguishing the wearer as someone not conforming to normative dress codes and thus susceptible to sanctions and interventions from his environment, against which, in turn, the design can protect.

Such protection comes at a price, as the bodyplasts constrict the body's movement and mould the skin/surface in an inflexible manner. Their flesh colour is not just a reference to the industrially made band-aid that is designed to blend with the organic surface; the plasters also press the body into amorphous (or even polymorphic) shapes. The synthetic bodyplasts-as-garments develop an autonomous surface perception. They do not cling to

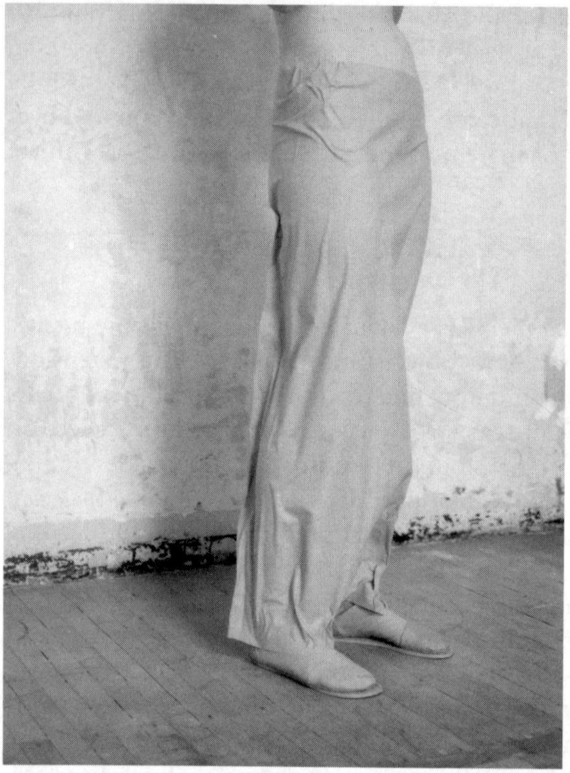

30 Carol Christian Poell, bodyplast long skirt, silk, leather socks with flat sole, 2000

the body's contours in an erotically charged manner, but rather present the wound as a structural transfer from negative space (as insertion) to artificial surface (as a positive layer of re-constitution).

Absence of surface

The fourth piece by Poell to be analysed for its surface value is a pair of transparent leather gloves that were made during the summer/autumn of 2001 (Figure 31). From Surrealist *objets trouvées* (the bronze cast in André Breton's novel *Nadja* of 1928) to Rita Hayworth's striptease in *Gilda* (1946) the removal of a dark glove and revelation of nude, white flesh has always operated as the *derobing* of sexuality: literally as the exposure of the skin/surface to present the body or its parts for consumption, metaphorically as the reification of the sexual act into a performance that can only be consumed in representation. The glove is the barrier between the apparent authenticity of the bare skin and the artifice of the garment, while constituting in fact the surface-object as simulacrum (habitually denoted as elegant and fashionable) of the sexualized female body.

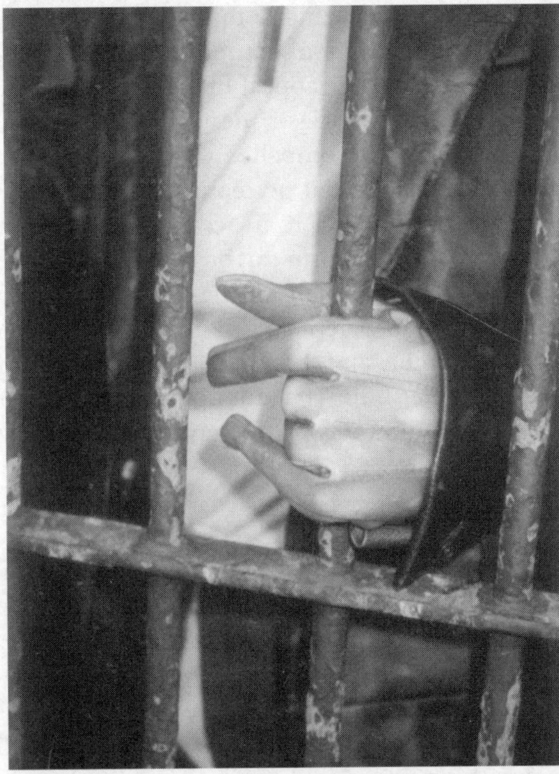

31 Carol Christian Poell, transparent leather gloves, 2001

However, when the glove is transparent and the revelation of the flesh thus presents itself without any performative interaction of the wearer, nakedness is simultaneously accentuated and removed. The man who is wearing the pigskin gloves – an object to be presently sexualized like Gilda?! – shows the colour and shape of his hand but, at the same time, conceals and processes it through a surface: the transparent leather fabric, its cut and visible stitching. The transparent glove is a material object but absents itself as a visual barrier. It is not a mere accessory but, through its dialectic of presence and absence, stages a debate about its own status. Does the glove exist as a fashion signifier when its designed function is to show its absence *as object*?

In modernity the absence or loss of the object has often been employed to underscore its meaning and cultural value. In Max Klinger's graphic cycle 'Paraphrase on the finding of a glove' (1881), as in *Nadja* some four decades later, the empty surface-cover of the hand stands for both the body that has been stripped naked and the replacement of the body as agent, as acting subject. The glove without a hand generates consternation, a feeling of loss, or even terror. Culture demands completing the object through the subject;

in Klinger's visual parable the woman who has lost the glove needs to be found, while in Breton's narrative the ideal Nadja is to be poetically evoked to match the artistic appeal of the sculptural glove.

When the surface of the glove, despite its object presence, is visually absent the gaze is compelled to flicker back to the signified of the hand to see whether the initial perception of a material surface can be verified. The shape of the glove has to be reconstructed mentally as an articulated layer in which the physical form can be clothed. This presence/absence of the surface becomes a concrete call for the simulacrum to be created (in this case, of a hand). The transparent (pig)skin/surface of the object displays its absence as a material value, and the simulacrum withholds the need to find an occupier for the leather hull. The autonomy of the object *as object* (not as a piece of design or commodity) is preserved here by its surface value.

Surface support

At one level of critical discourse fashion is just a façade, a screen that conceals the physical and material realities behind a surface of trend forecasting, styling and consumer profiles. The valorization of design occurs through commercial success, which is based on an insistence on *superficies* that are readily understood.

My final case study of Poell's work, a silicone-coated leather dress, makes this point quite literally (Figure 32). The sheet of fabric clings to the front of the figure, leaving the back bare, and thereby providing 'faciality' for one half of the body while leaving the rest bare and therefore unarticulated for fashion (even if sexualized as part of another discourse). The sur-face presents a minimal artifice on top of a corporeal front/face that clings as alien object to the skin.

The dress is not suspended from the usual points at the shoulder seam or waistline but requires a large surface area of the body as foundation. At the same time the dress declares its autonomy from the body: its silicon surface is the opposite to the skin/surface, in terms of texture and colour. The garment articulates a critical relationship between subject and object, between clothes that need the *physis* of the body to lend them their shape and those that retain independence from the body in order to fulfil their own object-value as commodities. The relationship between concealing and revealing, between form-giving and inanimate object, determines the impact of the garment as 'design' – understood in this context as a reified construction (faciality) that generates new surfaces for purposes of consumption.

The piece of silicone leather is not cut to a recognizable dress pattern or tailored into a garment. The shape that Poell uses to subvert the habitual

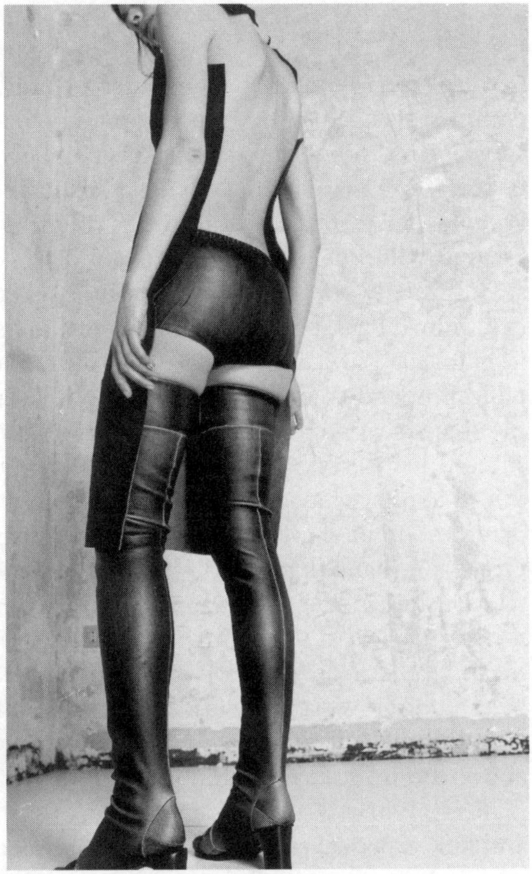

32 Carol Christian Poell, silicone leather stick-on dress, stretch leather toe-stockings with mid-heel sole, 2000

suspension of a garment from the body (shoulder seam, waist) is a simple rectangle with minimal alterations at the neckline. This abstraction removes it from any standard denotation as a 'dress' and establishes it as an object that, materially as well as structurally, is constituted by the dialectic between material and surface. Through its radical flat shape and the original 'stick-on' manner of wearing it, the piece is articulated first *as surface* – although imbued with distinct density and weight. Subsequently, its articulation *as object*, as dress, occurs through the female body that, significantly, does not lend it shape but supports it simply without compromising its own independence. The equation 'surface/material<->object' that establishes the dialectic between boundary and substance for a designed and articulated form here finds a concrete application within fashion whose inherent structure of reification thus allows again for a direct extension toward the equation of 'production/consumption<->object' that determines the use of an object as dress.

Conclusion

I stated at the outset that in order to understand the productive tension within surfaces they have to be seen as dialectic. The negation of material substances through the *superficies* is conflated critically with the affirmation of surface as material in order to constitute the object. And while Deleuze and Guattari concluded their work on faciality with the sentence 'le visage est une politique' ('the face is a politics'), which exposed concrete power structures at work,[18] my own present assessment of the 'sur'/'face' has been determined by the expansion of the 'fashion system' as a ubiquitously coded layer for such structures. Surfaces expand and multiply within such systems as reification and commodification take hold, not only as a binary opposition to subjectification but also as a flickering back-and-forth between objects and subjects, which mutually constitute themselves anew at every cultural turn.

As a surface moves in its characterization from boundary – an ontological demarcation of empty space – toward an inscribed, coded and commodified projection plane that reflects the values that instrument power relations, its relationship to the object/substance changes. This shift entails a constant flickering back to the signified of the object, in order to comprehend the meaning that is borne on its instantly accessible surface. The consumption of the object as basis for its production here occupies a level of representation beyond mere 'faciality'. The 'sur' in 'surface' is the concrete elevation above the front/face of the 'fashion system' into a reading of surfaces as material. Whereas such understanding of surfaces might usher in innovation and new experiences in design, the objectified body in fashion cannot avoid reaching herein its critical apotheosis as material for designed surfaces and surface for desired materiality.

Acknowledgements

Some of the material analyses in this essay were conceived originally between 2007 and 2009 for the Carol Christian Poell Spezialarchiv in the Museum for Applied Arts (MAK) in Vienna; I am very grateful for its permission to re-write these paragraphs for the present publication.

Notes

1 Theodor W. Adorno formulated in his exile in California (1944–47): 'Es gibt kein richtiges Leben im falschen' ('There is no right life within false one' or 'Wrong life cannot be lived rightly'), 'Minima morialia' 1, 18, in *Gesammelte Schriften*, vol. 4 (Frankfurt-on-Main: Suhrkamp, 2003), p. 19 – individual, subjective morality cannot become possible within structural falsehood of an alienating and commodified social system.

2 *Le Grand Robert de la langue française* (2nd edn) (Paris: Le Robert éditions, 2001), p. 873.

3 *The Oxford English Dictionary* (2nd edn) (Oxford: Clarendon Press, 1989), p. 286.

4 In English in the original text.

5 K. Marx, *Grundrisse der Kritik der politischen Ökonomie (Rohentwurf), 1857–1858* (2nd edn) (Berlin: Dietz, 1974), p. 13; translated by M. Nicolaus as *Grundrisse: Foundations of the Critique of Political Economy (Rough Draft)* (London: Allen Lane, 1973), p. 91 (translation modified).

6 See G. Simmel, 'Die Mode: zur philosophischen Psychologie', in *Philosophische Kultur: Gesammelte Essais* (Leipzig: Klinkhardt, 1911), pp. 39–40.

7 Deleuze would characterize later the methodology of the chapter on 'faciality' in *Mille plateaux* as 'lignes de fuite' (lines of flight); see G. Deleuze, *Pourparlers* (Paris: Éditions de Minuit, 1990), p. 233.

8 G. Deleuze and F. Guattari, *Capitalisme et schizophrénie*, vol. 2, *Mille plateaux* (Paris: Éditions de Minuit, 1980), pp. 205–208; translated by B. Massumi as *A Thousand Plateaus: Capitalism and Schizophrenia* (London: Athlone, 1988), pp. 167–170.

9 Deleuze and Guattari, *Mille plateaux*, p. 208; Deleuze and Guattari, *A Thousand Plateaus*, p. 170.

10 See N. Abraham, *L'Écorce et le noyau* (Paris: Aubier/Flammarion, 1978), pp. 203–226 (this article, that lends its name to the entire volume by Abraham, was originally published in *Critique*, February 1968); and D. Anzieu, *Le Moi-peau* (Paris: Bordas, 1985); translated by C. Turner as *The Skin Ego* (New Haven and London: Yale University Press, 1989), pp. 9, 40, 62–65 *et passim*.

11 Deleuze and Guattari, *Mille plateaux*, p. 222; Deleuze and Guattari, *A Thousand Plateaus*, p. 181.

12 Obviously, patterns can be imprinted or embroidered *onto* a surface as well as being woven or dyed *into* the surface during production.

13 They do this 'collectively', no less, thereby contradicting the repeated claims of structural innovation in fashion and, again, the possibility of actual subjectification within it.

14 'Pieces' is the term Poell uses exclusively to define his work, with the immediate caveat that he refutes any categorization of his designs as 'art'.

15 On a semiotic level this technique, as 'pli selon pli', had already been pioneered by Stéphane Mallarmé's poetic layers in the latter half of the nineteenth century; Pierre Boulez employed its structure serially in his eponymous composition of 1957–62.

16 As dramatically depicted in the filmic documentation *The Action Art of Hermann Nitsch from Past to Present* (Berlin: Kröthenhayn, 2006).

17 The design might also be ritualizing a menstrual pattern, in traces of blood depicted as running down the legs of female trousers?!

18 The full passage reads: 'The order is totally different: despotic and authoritarian concrete assemblage of power —> triggering of the abstract machine of facilitiality [*visagéité*], white wall/black hole —> installation of a new semiotic of signifiance and subjectification on that holey surface … The face is a politics'. Deleuze and Guattari, *Mille plateaux*, p. 222; Deleuze and Guattari, *A Thousand Plateaus*, p. 181.

PART III

DEPTH, IM/PERMEABILITY

13 ✧ Surface recognition: light and reflection in mid-nineteenth-century drawing rooms

Alice Barnaby

Nay, if there's room for poets in this world
A little overgrown (I think there is)
Their sole work is to represent the age
Their age, not Charlemagne's, – this live, throbbing age,
That brawls, cheats, maddens, calculates, aspires,
And spends more passion, more heroic heat,
Betwixt the mirrors of its drawing-rooms,
Than Roland with his knights at Roncesvalles.
To flinch from modern varnish, coat or flounce,
Cry out for togas and the picturesque,
Is fatal, – foolish too.

(Barrett Browning, 1856: 200–210)

In her gender – and genre-defying novel-poem *Aurora Leigh*, Elizabeth Barrett Browning positively sets contemporary fashions for the female, socially intimate space of the nineteenth-century drawing room against and above the more culturally elite appetite for idealized representations of masculine medieval chivalry. Barrett Browning rejects the period's love affair with the poetic romances of Tennyson and the paintings of the Pre-Raphaelite Brotherhood and recognizes and embraces instead the passionate yet problematic features of her modern age, embodied in this passage by the stylishly commodified space of the drawing room. In the space of eleven lines of blank verse she unsettles the dominant discourse of separately gendered spheres and warns against the impulse to retreat in distaste from new styles of mirrors, varnished furniture and draperies. For Barrett Browning then, the drawing room is neither a passive nor discretely private space, but one which is intrinsically, and to a certain extent heroically, connected with the nineteenth-century pulse of modernity. Barrett Browning gives voice to a realistic rather than romantic heroism, which is crucially also a feminine heroism, as much about the materiality of urban existence as about literary intellectual activity.

The vision of a bold, energetic and outward-facing partnership between drawing room and female subjectivity is at odds with more established interpretations of Victorian life, which tend to frame both as elements in retreat from public life.[1] Yet, as this essay will suggest, and as Barrett Browning herself intimates, a significant relationship *did* exist between the emerging practice of professional interior design, and modern female self-determination and creative expression. Of particular interest is the fact that Barrett Browning goes further than merely identifying an active female space: she also draws attention to specific aesthetic markers of the drawing room – the surfaces of mirrors, polished furniture and luxury fabrics. It is across these reflecting and refracting surfaces that we can catch glimpses of the extent to which nineteenth-century lives were subtly affirmed or challenged by their reflections in material surroundings.

Many years after the publication of *Aurora Leigh*, Virginia Woolf acknowledged the heroism and practical wisdom of Barrett Browning's words: 'We may suspect that Elizabeth Barrett was inspired by a flash of true genius when she rushed into the drawing-room and said that here, where we live and work, is the true place for the poet.'[2] This essay does not propose that all nineteenth-century drawing rooms and their surfaces were spaces of poetry, but it does suggest that drawing rooms were frequently spaces of female creativity and identity formation. We can see from the example in *Aurora Leigh* that the stylish surfaces of the drawing room were in no way neutral – Barrett Browning tactically uses the contemporary taste for gleaming polish and glistening mirrors to pose an intellectual and gendered challenge to the authority of the male artistic voice of the period. With this in mind it is perhaps no coincidence that these surface markers of femininity and feminine space were held in balance between other historically defined pressures relating to judgements about quality: the imperative to define something or someone as refined or vulgar; passive or active; trivial or serious. These dyads of discrimination were also criss-crossed and complicated by further assumptions concerning social class. So, although an unusual topic for poetic analysis, attention to surface details of interior design was a much discussed and judged expression of personal taste in areas of household design, luxury commodities and matters of etiquette.

After taking the initial cue from Barrett Browning this essay will move on to explore how a variety of commentators, such as Charles Dickens, John Ruskin and Edgar Allan Poe among others, all recognized that an aesthetics of surface stretched across the social space of the drawing room, both reflecting and stimulating vigorous debates about taste, class, gender and domestic space. This was a period in which surfaces themselves were commodified and started to function as ways of projecting, augmenting, shaping, revealing or disguising individual and group subjectivities. But these reflecting surfaces are only one side of a perceptual partnership; for without light there would be no effects of gleam, shine or lustre. It is no coincidence that as artificial illumination in the nineteenth century became brighter, cheaper and more varied, surface styles intimately responded to and exploited the properties specific to candle, oil or gaslight.[3] The essay will also consider, therefore, how gaslight in particular presented both new possibilities and new problems for the aesthetic status of domestic surfaces. Seemingly superficial matters such as glittering surface reflectance or the deceptively deep lustre of veneer could provoke approval and aspiration as well as anxiety about threats to social hierarchy and a decline in social standards. By attending to the historical significance of surfaces we can discern how identities are in part dependent upon the extent to which our environment allows us to recognize ourselves and each other in its surfaces. Towards the end of this essay I indicate how this interpretation of surfaces is influenced by recent studies in theories of enactive cognition which locate cognitive agency in the environment as well as the human form.

For centuries, warm and glowing effects had been coaxed from dull wooden furniture through the transformative arts of varnishing and japanning.[4] But by the end of the eighteenth century attention to aesthetic interactions between illumination and domestic surfaces intensified, resulting in a proliferation of manuals, techniques and materials. The definition of polishing in a domestic encyclopaedia of 1803 explains the process as an 'act of smoothening [*sic*] and imparting brightness to hard substances, such as metals, marble, glass, etc. by rubbing them with certain matters adapted to the purpose'.[5] In this interpretation lighting effects were bestowed upon substances by way of human action rather than being drawn out from their inherent material nature. Further artificial manipulation occurred when combinations of these materials were juxtaposed, resulting in a complex variety of reflective surfaces as is seen in Plate 28, which shows a torchère candelabrum made by the influential cabinet-maker George Bullock between 1816 and 1818. Illumination from candles was increased and subtly altered by the juxtaposition of glass, metal and oak veneer combined with the effects of ebony and gilt gesso. Polishing these various materials created a partnership between the raw substance and the skilful act of its refinement. Treating furniture in this manner was of particular interest to a late eighteenth-century French visitor to Britain whose travel account noted that 'their tables and chairs are also made of mahogany of fine quality and have a brilliant polish like that of finely tempered steel'.[6] Such items were costly to produce and even costlier to purchase. Bullock claimed some socially significant clients such as Sir Walter Scott whose belief in Bullock's 'uncommon taste and talent' prompted him to engage the cabinet-maker for a large commission at his home Abbotsford near Melrose in Scotland.[7] Yet as the nineteenth century progressed the allure (and accessibility) of highly polished surfaces was not just restricted to an elite clientele.

In the early nineteenth century English cabinet-makers began to practise the technique known as 'French polishing' which, through considerable manual labour and skill, produced stunning effects of chatoyement or changeable lustre. The first reference to the method in *The Times* occurred in 1824 in an advertisement placed by a young man looking for a position as a cabinet-maker who stated that he was familiar with the practice of 'French Polishing' ('To Cabinet-Makers'). Two years later *The Practical Cabinet-Maker and Upholsterer* by Peter Nicholson included an entry on the subject in the glossary:

> French polish is a new and durable mode of polishing or varnishing, by which means it is not so much necessary to polish the surface of the wood itself. This mode consists in applying a considerable thickness of *transparent* gum-lac over the surface, so that the surface of gum appears as if it were the surface of the wood.[8]

Through this specialist and non-functional aesthetic procedure articles of furniture were transformed into eye-catching, luminous surfaces that exploited any available ambient light. In the same year as the advertisement in *The Times*, J. W. Neil published *The Painter's Guide to the Art of Varnishing and Polishing* which advised that gum copal was the superior type of varnish because 'it exceeds all others, possessing perfect transparency, limpidness, and lustre with a sufficient body, which serves as a solid glazing'.[9] This practice could almost be likened to a form of alchemy whereby the craftsman's skilled labour turned a piece of opaque wood into an object that was capable of performing small but complex illuminated displays.

Beyond the sphere of cabinet production, early nineteenth-century experts in matters of domestic management favoured mahogany above all other woods as it did not require professional treatment in order to produce an effect of high shine. In 1825 the author of *Domestic Duties* explained that 'mahogany has the most valuable qualities ... it is capable of receiving, by mere friction, the highest polish'.[10] The advice lists further pieces of household equipment which through physical treatment could achieve an aesthetic as well as functional status within the home. The author details how 'Bright stoves, highly-polished fire-irons, and steel fenders, are amongst the indispensable parts of drawing-room furniture. These should be rubbed every day with a leather, to preserve the polish.'[11] Evidently, functional objects as well as luxury objects were valued for their capacity to participate in the domestic display offered by the partnership of light and surface.

However, intimations of an impending crisis about the morality of aesthetic developments in household design had been on the cultural horizon for some time before the middle of the nineteenth century. In 1800 James Roberts's manual for the polite domestic pastime of transparent painting warned genteel readers that aesthetic errors of judgement had wider, damaging implications for one's cultural credibility: 'I do not advise a scholar to varnish too highly; it spoils the delicacy of the atmosphere, and gives a picture the appearance of a tea-tray … or a Birmingham sign.'[12] Roberts's reference to a tea-tray and to Birmingham carries with it a whole set of associations concerned with the commerce of new luxury, but not necessarily quality, objects. As Maxine Berg has detailed in *Luxury and Pleasure in Eighteenth-Century Britain*, from the middle of the eighteenth century Birmingham grew as a leading centre in the production of consumer goods and acquired a reputation for specializing in inexpensive, but nevertheless fashionable, consumer objects. Less valuable materials of steel, brass and copper and new surface techniques of plating, casting, polishing and gilding were used in preference to more costly and traditional metals of solid gold – or silver.[13] These new methods and product ranges challenged established London-based businesses of gold and silversmiths, provoking heated debates about quality control, commercial aspirations and standards of taste.

By the 1820s voices critical of the pressure of popular fashion upon established standards of taste were becoming increasingly strident. In *The Grecian, Roman and Gothic Architecture*, William Fox Jr, son of the merchant and radical pamphleteer William Fox, lamented a paucity of quality in interior design:

> The bad taste displayed in the interior of many of our best houses is also exceedingly striking – rooms are crowded with superfluous furniture … the eye, instead of being gratified with an assemblage of what is beautiful and chaste, and *appropriate*, is repelled by the glare of a mass of whimsical and costly absurdities.[14]

This evaluation of taste includes negative assessments of an inability to judge both quantity and quality. According to Fox the viewer is assaulted with unnecessary numbers of objects coupled with glaring effect, which rather than seducing and pleasing the eye, in fact initiate a gap between subject and object. The perimeters of this gap are defined by an opposition between the corruption of the senses on one side and the rigours of the mind on the other. The writer aligns glaring objects with states of irrationality through references to absurdity and unreliable whimsy. Good taste was ruled by the mind and supported by the senses whereas bad taste seduced the senses and mental principles were abandoned. William Hazlitt pursued similar aesthetic concerns to Fox when the same year he wrote his essay 'On Vulgarity and Affectation' in which he maintained that surface appearance was not a trustworthy or reliable indicator of deeper aesthetic sensibilities:

> I would rather see the feelings of our common nature (for they are the same at bottom) expressed in the most naked and unqualified way, than see every feeling of our nature suppressed, stifled, hermetically sealed under the smooth, cold, *glittering varnish of pretended refinement and conventional politeness* [emphasis added].[15]

By conflating the properties of animate and inanimate substances Hazlitt draws attention to the interdependent nature of identity, which is formed through the relationship between individuals and their property. The layering technique of French polishing with its 'glittering varnish' is co-opted into an attack on certain personal values and behaviour. It as if people have become furniture and their purity of 'nature' is in danger of being suffocated by an artifice of false appearance.

Hazlitt was not alone in this particular mode of expressing such preoccupations. Throughout his novels Charles Dickens blurred the boundaries between human and in-human to great comic effect. In *Little Dorrit* (1855–57) when describing the character of the widow Mrs General, Dickens (like Hazlitt) relies on the craftsman's technique of varnishing to construct a metaphor of human flaws. Under Mr Dorrit's employ Mrs General is required to 'finish' or polish off the rough corners of his daughters' social etiquette. The narrator observes that 'The more cracked it was, the more Mrs General varnished it. There was varnish in Mrs General's voice, varnish in Mrs General's touch, an atmosphere of varnish round Mrs General's figure.'[16] Rather than attempting to heal the cause of a problem, Mrs General, like varnish, resorts to a quick fix, covering up and disguising fractures symptomatic of a deeper malaise. The highly reflective finish of varnish has become an active, unpleasant and unboundaried agent seeping out from Mrs General's body, infecting her speech, extending her spatial presence in a hard, clear, shiny aura and, like King Midas, transforming everything that comes into contact with her.

Discussion about the relative merits of polished surfaces was also to be found in the relatively new discourse and discipline of interior design. Charles Eastlake's seminal text *Hints on Household Taste in Furniture, Upholstery and Other Details* (1868) paid close and judgmental attention to the minutiae of new domestic styles. On the subject of wood carvings he stated that:

> Whatever the faults of the modern treatment may be, they are rendered doubly objectionable by the application of varnish. The moment a carved or

sculptured surface begins to *shine*, it loses interest. But machine-made ornament, invested with an artificial lustre, is an artistic enormity which should be universally discouraged.[17]

The message is clear: commerce has insinuated itself into the domestic environment tainting it with an unnatural finish. Artistic integrity is lost and the thought or 'interest' which previously inspired creativity has been superseded by 'machine-made' products coated in thoughtless surfaces, empty of meaning and asethetic value.

It was not just the technique of polishing surfaces that came under attack; other treatments such as veneering and gilding were also targets for arbiters of taste and craftsmen alike. The circulation of instruction manuals such as Whitaker's *The House Furnishing Assistant* provided professional guidance for tradesmen keen to satisfy the whims of demanding and fashion-conscious clientele. Whitaker warns that those who:

> look to have their money's worth in material or labour, quite indifferent as regards artistic character or merit of design will suffer in that … So soon as time has destroyed the varnish, lustre, polish, or gliding upon which alone the attraction of the article depended, the value will depart with its adventitious causes.[18]

The seductive surfaces that exploited and manipulated light were not reliable measures of worth and were not substitutes for labour and enduring material. What lies on the surface can only temporarily deceive the senses and will eventually betray poor quality goods and craftsmanship. In the same publication the glossary entry for gilding highlights the importance of exercising moderation: 'the parts to be gilt should be chosen with care, and not over abundant, which would produce a tawdry effect'. In the entry under 'V', veneering is described as a process that makes the 'surface appear as if the material had been of the finest and most rich quality; for this reason the ground may be of inferior wood, and the veneer of the most superior kind'.[19]

The potential pitfalls of decorating one's home in this manner are played out in Dickens's portrayal of Hamilton and Anastasia Veneering in *Our Mutual Friend* (1864–65). Dickens's authorial voice takes a negative stance towards the couple's taste for interior design and newness which is ridiculed in both their name and an episode that exposes their parvenu social aspirations:

> For, in the Veneering establishment, from the hall-chairs with the new coat of arms, to the grand piano-forte with the new action, and upstairs again to the new fire-escape, all things were in a state of high varnish and polish. And what was observable in the furniture, was observable in the Veneerings – the surface smelt a little too much of the workshop and was a trifle sticky.[20]

Rather than possessing an effortlessly and inherently smooth sociability, the Veneerings experience a 'new' and 'sticky' status, recently bought like their furniture from a workshop. Not content with using the surface qualities of polished wood to signify new wealth's drive for cultural status, Dickens proceeds to underscore the commentary with an inventory of other domestic objects and the attributes of the guests who are captured in a complex web of reflections, 'The great looking-glass above the sideboard, reflects the table and the company. Reflects the new Veneering crest, in gold and eke in silver, frosted and also thawed, a camel of all work.'[21] Like the furniture, Dickens's Veneerings were made from a material whose 'inferior' substance was concealed by the deceptive surface appearance. Dickens pinpoints a palpable tension between the ephemeral nature of these reflecting surfaces and their owners' vigorous struggle to attain symbolic capital and enduring social recognition.

Negative judgements regarding the commodification of gleaming and glittering luxury items, and their betrayal of class stratifications and insecurities within the new middle classes, were exacerbated by the increased accessibility of gas lighting.[22] By the middle decades of the century many upper – and middle-class domestic interiors were equipped with new lighting fixtures and fittings that produced a considerably brighter source of illumination, further animating and augmenting the play of reflected light. In the 'Philosophy of Furniture' (1840) Edgar Allan Poe launched a withering attack on the Anglo-American use of gas within domestic settings. He was particularly exercised by the taste for glittering and glaring effects which accompanied artificial gas lighting. The essay presents an unlikely unification of two traditionally distant and hierarchically opposed forms, the intellectual practice of philosophy and the lowly artisanal practice of furniture making. Through this coupling Poe implied that furniture had the capacity to think and that philosophy could be a tangibly felt presence.

> *Glare* is a leading error in the philosophy of American household decoration — an error easily recognised as deduced from the perversion of taste just specified. We are violently enamoured of gas and of glass. The former is totally inadmissible within doors. Its harsh and unsteady light offends. No one having both brains and eyes will use it … It is not too much to say, that the deliberate employer of a cut-glass shade, is either radically deficient in taste, or blindly subservient to the caprices of fashion. The light proceeding from one of these gaudy abominations is unequal, broken, and painful.[23]

On the subject of mirrors and other reflecting surfaces Poe complained that with 'the attendant glitter upon glitter, we have a perfect farrago of discordant and displeasing effects'.[24] His concern about sensory over-stimulation and his aesthetic antipathy toward the flickering, glittering and glaring domestic light marked a point of rejection and repulsion against the proliferation of highly ornamented illuminations. This was more than just a critique of style, however, because for Poe the new domestic environment challenged his place and participation in the structure of social intercourse. In effect Poe's self-recognition in the social space of the drawing room had ceased to be coherent and intact, but had instead become nightmarishly multiplied and fragmented due to the blinding effects of garish gaslight. Poe was displeased with the wider significance of what these reflections represented and he set himself apart from the offensive and tasteless 'caprices of fashion', not wishing to share his own reflection with others in the drawing room's modern partnership of light and surface.

Belief that perceptual attraction towards the lighting effects of glare and glitter was detrimental to the cultivation of good taste was frequently accompanied by a belief that morality was also at stake. For example, in an article from the *Ladies Repository* entitled 'The Moral Worth and Beauty of Religion', the Reverend Uriah Clark maintained that 'Wealth and splendour may fascinate the eye, but can impart no principle to the mind.' He worried that because 'Multitudes have been carried away by these glittering baubles … Virtue has been tarnished by the glitter of silver.'[25] Like Poe the Rev. Clark constructed a negative association between pleasure, perception and certain lighting practices in which he warned readers about the dangers of 'the glitter of sensual pleasures'.[26] Here, in this publication aimed at a middle-class female readership, we can see that an implicitly gendered interplay between inanimate surface qualities and animate subjectivities emerged. And because notions of femininity were constructed through an emphasis on surface rather than depth it therefore followed that anxieties about wayward female virtue were frequently coupled with aesthetic judgements regarding surface qualities.

Seven years later the twinned concerns of the gendering of taste and the nation's domestic moral health reached a visual climax in William Holman Hunt's painting *The Awakening Conscience* (1853) and John Ruskin's response in *The Times* to that image. The painting (Plate 29) depicts a woman located in a drawing room caught in a state of acute moral realization when she is shown on the point of moving physically and emotionally away from her lover. Her gaze is focused upon and lit by the direct presence of daylight which floods in from the window. Domestic objects of luxury that display dubious surface qualities of varnish, polish, veneer, reflection and gilding surround the woman who, the viewer infers, was bought just like the ornaments; as the art historian Caroline Arscott says, 'The items are newly purchased and installed as is she.'[27] According to Ruskin these objects were 'common, modern, vulgar' and with respect to the overall interior and all that it stood for he asked 'is there nothing to be learnt from that terrible lustre of it, from its fatal newness?'[28] Eastlake shared Ruskin's abhorrence of such modern materiality and the popular appetite for 'newness' which he believed women were particularly susceptible to: he quipped that 'the ladies like it best when it comes like a new toy from the shop, fresh with recent varnish and untarnished gilding'.[29]

Whilst the commodities in Hunt's painting produce a complex and captivating manipulation of light, the woman is represented in a moment of transition in which she moves away from the abundantly sensuous environment towards the unprocessed and unmediated light of the day. Thus two different kinds of light and accompanying visuality are set at odds with each other, thereby exemplifying the cultural anxieties that surrounded expressions of status, value, gender and morality. Rosemary Betterton observes that 'The immorality of the gentleman and his mistress has infected the inanimate objects which surround them.'[30] However, infecting sensory activity flows back *and* forth between subject and object, rather than remaining limited to the subject's influence upon an object. Suggesting that the inanimate surface of objects possesses independent agency or intentionality might at first seem somewhat far-fetched, yet these surfaces evidently had a strong perceptual effect provoking both attraction and repulsion.

Women's involvement with the domestic decorative surfaces of their homes was not always, however, to their detriment. Their identity was not always negatively aligned with connotations of inappropriate luxury, immoral behaviour or the mistress's passive lifestyle as represented in *The Awakening Conscience.* In fact in the latter half of the nineteenth century women took an increasingly professional interest in the design and decoration of their own and others' living spaces. Two principal agents in this shift were the cousins Agnes and Rhoda Garrett whose publication *Suggestions for House Decoration in Painting, Woodwork, and Furniture* (1876) provided readers with precise guidance on matters of household taste. As with other commentators, surface appearance continued to be a major concern for these writers who, in their introduction, explained:

> It will be enough for our purpose if we can help to make it understood wherein lies the difference between good work and bad; between wall decorations chosen at random from an upholsterer's pattern book, and those designed or selected by an eye trained to discern the subtleties of shade and colour, or accustomed to judge between frivolous designs and bad construction, and simple straightforward forms which do not attempt to conceal by a mass of senseless ornament and a misleading shimmer of French polish.[31]

While 'misleading shimmer' was frowned upon as an indication of poor aesthetic judgement the cousins did not condemn all light-catching surfaces in the home. Beeswax and turpentine were recommended as aids to enhancing the glow of furniture over time, and a mirrored cabinet in

American walnut was promoted for its capacity to display china and 'give a brilliant effect to the whole'.[32] As far as metal work was concerned a sense of the value of time was also involved in the Garretts' advice. New highly ornamental implements such as fenders and fire irons were disregarded in favour of refurbishing older brass and steel household tools, for according to these professional designers 'Nothing better than polished brass can be used in the manufacture of these articles ... upon which skill as well as time and money have been expended.'[33] Publications such as the Garretts' demonstrate that women had an increasing amount of aesthetic control over their environments but also that surface qualities continued to trigger strong cultural responses about the formation of taste.

What then was the sensory appeal (and anxiety) about living among such light-sensitive possessions whose surfaces gleamed and shone so vividly in response to ambient lighting conditions? Perhaps one answer lies in the specific nature of the relationship that was established between person and the shining possession. A reflective surface always acknowledges your presence and triggers a kind of non-linguistic dialogue. It is unwaveringly attentive, producing a reassuring glimmer of aesthetic recognition as you enter and move through a room. As the surface tracks and reacts to your movement it takes on a lively quality of vitality and agency. In so doing it has the somewhat uncanny capacity to look back and objectify you, the subject. As Jacques Lacan suggests, such a partnership between light and surface can either affirm or threaten a sense of self. He explains:

> that which is light looks at me, and by means of that light in the depths of my eye, something is painted – something that is not simply a constructed relation, the object on which the philosopher lingers – but something that is an impression, the shimmering of a surface ... which is no way mastered by me.[34]

Lacan proposes that a deep psychological anxiety about an individual's loss of reason or control occurs when encountering him – or herself reflected and therefore 'othered' in 'the shimmering of a surface'. With that in mind I suggest that the influence of surface behaviour upon self-consciousness operates at both an individual and collective level. It therefore follows that different aspects of cultural confidence can be either confirmed or contested due to anxieties about the status of who is being reflected. In this way reflections also afford a form of Hegelian recognition which is both a political and aesthetic act of consciousness. Axel Honneth's more recent work in the philosophical ethics of recognition goes further in underscoring the importance of subject-object relations in the construction of self-image:

> In perceiving one's own vocal gesture and reacting to myself as my counterpart does, I take on a decentred perspective, from which I can form an image

of myself and thereby come to a consciousness of my identity … individuals can only become conscious of themselves in the object-position.[35]

If we substitute 'vocal' with 'visual' gesture it then becomes possible to gauge the significance of domestic reflective surfaces in their capacity to objectify *and* produce a sense of identity. Reference to Hegel and Honneth might seem slightly overblown for the everyday context of the drawing room, but as new nineteenth-century techniques developed and the accessibility and affordability of luxury commodities increased, there was also a notable increase in voices articulating their concern about the propriety of owning such objects. This concern centred on the fact that many commentators disliked what they saw in society's reflection; they were threatened by who was demanding aesthetic recognition. Domestic reflections had the capacity to trigger consciousness of both class and gender identities.

The premise that our surroundings stimulate acts of recognition and thought processes can be taken one step further for it seems that drawing room reflections were actually ephemeral acts of consciousness taking place beyond the boundaries of the subject's body. From the fields of Cognitive Sciences and Science and Technology Studies there is firm theoretical evidence to support such a proposition. In particular, recent theories of enactive cognition offer an alternative to the standard Cartesian or brainbound model of cognition.[36] The philosopher Andy Clark believes that thinking exceeds the mind and occurs in a sort of feedback loop between mind, body and world. He claims:

> It matters that we recognize the very large extent to which individual human thought and reason are not activities that occur solely in the brain or even solely within the organismic skin-bag. This matters because it drives home the degree to which environmental engineering is also self-engineering. In building our physical and social worlds, we build (or rather, we massively reconfigure) our minds and our capacities of thought and reason.[37]

Clark's phrasing here is striking, and perhaps a little odd, particularly his term 'organismic skin-bag' to describe the surface of the human body. Nevertheless this unfamiliar disciplinary world and its equally unfamiliar language offer new possibilities for the cultural historian because of the way in which Clark values the role of the environment in the formation of a thinking subject. He believes that the act of thinking takes place out there in the world, rather than remaining purely within the brain and the body. We should therefore not underestimate the extent to which the textures of our physical world contribute to our sense of self within society. It is perhaps then no surprise that commentators such as Barrett Browning, Ruskin, Eastlake, Dickens, Poe and Holman Hunt recognized the pressures of modernity by interpreting society's reflections in the surfaces of domestic surroundings. During the middle decades of the nineteenth century however materially thin surfaces might have been, they were in no way culturally or aesthetically insignificant; these surface reflections possessed a unique capacity to confer or deny cultural recognition.

Notes

1 Since Leonore Davidoff and Catherine Hall's groundbreaking work in *Family Fortunes: Men and Women of the English Middle Class 1780–1850* (London: Hutchinson,1987),a number of scholars have been developing and productively complicating this structure. See for example Sharon Marcus's *Apartment Stories: City and Home in Nineteenth-Century Paris and London* (Berkeley: University of California Press, 1999) and Amanda Vickery's *Behind Closed Doors: At Home in Georgian England* (New Haven: Yale University Press, 2009).

2 V. Woolf, 'Aurora Leigh', *The Second Common Reader* (A. McNeillie ed.) (New York: Harcourt Brace Jovanovich, 1986), p. 213.

3 For a fuller discussion of the cultural significance of light in the nineteenth century see the following studies: A. Barnaby, 'Light touches: cultural practices of illumination, London 1780–1840', PhD Thesis, University of Exeter, 2009; L. Nead, *Victorian Babylon: People, Streets and Images in Nineteenth-Century London* (New Haven: Yale University Press, 2000); C. Otter, *The Victorian Eye: A Political History of Light and Vision in Britain, 1800–1910* (Chicago: University of Chicago Press, 2008); W. Schivelbusch, *Disenchanted Night: The Industrialization of Light in the Nineteenth Century* (Berkeley: University of California Press, 1995).

4 Take for example John Stalker's treatise on the subject, *A Treatise on Japanning and Varnishing* (London, 1688).

5 A. F. M. Willich, *The Domestic Encyclopaedia; Or, A Dictionary of Facts, and Useful Knowledge. Comprehending a Concise View of the Latest Discoveries, Inventions, and Improvements, Chiefly Applicable to Rural and Domestic Economy: Together with Descriptions of the Most Interesting Objects of Nature and Art* (London: Murray and Hingley, 1803), p. 304.

6 F. de la Rouchefoucauld, *A Frenchman in England: 1784* (S. C. Roberts trans.) (Cambridge: Cambridge University Press, 1933), p. 42.

7 W. Scott, *Memoirs of the Life of Sir Walter Scott* (John Gibson Lockhart ed.) (Paris: Baudry, 1837). Letter Abbotsford 4 May 1818, p. 275 (Google Books, accessed March 2011).Other large commissions for Bullock included Great Tew Park and Cholmondely Castle.

8 P. Nicholson, *The Practical Cabinet-Maker and Upholsterer and Complete Decorator* (London, 1826), p. 7.

9 J. W. Neil, *The Painter's Guide to the Art of Varnishing and Polishing* (London: W. Davy, 1824), p. 39.

10 W. Parkes and F. B. Parkes, *Domestic Duties: Or, Instructions to Young Married Ladies, on the Management of Their Households, and the Regulation of Their Conduct in the Various Relations and Duties of Married Life* (London, 1825), p. 198.

11 *Ibid.*, p. 200.

12 J. Roberts, *Introductory Lessons, with Familiar Examples in Landscape, for the Use of*

Those who are Desirous of Gaining Some Knowledge ... of Painting in Water colours ... To Which are Added, Instructions for Executing Transparencies, in a Style both Novel and Easy (London: W. Bulmer, 1800), p. 33.

13 M. Berg, *Luxury and Pleasure in Eighteenth-Century Britain* (Oxford: Oxford University Press, 2005), pp. 159, 167.

14 W. Fox, *The Grecian, Roman and Gothic Architecture Considered as Applicable to Public and Private Building* (London: John Hatchard and Son, 1821), p. viii.

15 W. Hazlitt, 'On Vulgarity and Affectation', in *Table-talk: or, Original Essays*, vol. 2 (London: John Warren, 1821), p. 391.

16 C. Dickens, *Little Dorrit* (Oxford: Oxford University Press, reissued 2008), p. 377.

17 C. Eastlake, *Hints on Household Taste in Furniture, Upholstery and Other Details* (London: Longmans & Co., 1868), p. 85.

18 H. Whitaker, *The House Furnishing Assistant: The Practical Cabinet-Maker and Upholsterer's Treasury of Design* (London: Peter Jackson, late Fisher, Sons, & Co., 1847), p. 3.

19 *Ibid.*, p. 8.

20 C. Dickens, *Our Mutual Friend* (Oxford: Oxford University Press, reissued 2008), p. 6.

21 *Ibid.*, pp. 20–21.

22 S. Milan, 'Refracting the gaselier: understanding Victorian responses to domestic gas lighting', in I. Bryden and J. Floyd (eds), *Domestic Space: Reading the Nineteenth-century Interior* (Manchester: Manchester University Press, 1999), pp. 84–102.

23 E. A. Poe, 'Philosophy of Furniture' (1840), in *The Works of Edgar Allan Poe*, vol. 5 (Teddington: Echo Library, 2007), p. 2.

24 *Ibid.*, p. 4.

25 Rev. U. Clark, 'The Moral Worth and Beauty of Religion', *Ladies Repository*, 14 (1846), 354–356.

26 *Ibid.*, 356.

27 C. Arscott, 'Employer, husband, spectator: Thomas Fairburn's commission of *The Awakening Conscience*', in J. Wolff and J. Seed (eds), *The Culture of Capital: Art, Power and the Nineteenth-Century Middle Class* (Manchester: Manchester University Press, 1988), p. 170.

28 J. Ruskin, 'The Pre-Raphaelites', *The Times*, 25 May 1854, 7, *Times* archive, accessed 12 December 2008.

29 Eastlake, *Hints on Household Taste*, p. 83.

30 R. Betterton, *Looking On: Images of Femininity in the Visual Arts and Media* (Abingdon: Taylor & Francis, 1987), p. 89.

31 R. Garrett and A. Garrett, *Suggestions for House Decoration in Painting, Woodwork, and Furniture* (London, 1876), pp. 33–34. Other texts by female writers and designers include: L. Orrinsmith, *The Drawing Room and its Decoration and Furniture* (London: Macmillan, 1876); M. A. Barker, *The Bedroom and the Boudoir* (London: Macmillan, 1878).

32 Garrett and Garrett, *Suggestions for House Decoration*, p. 62.

33 *Ibid.*, p. 66.

34 J. Lacan, *The Four Fundamental Concepts of Psychoanalysis* (Alan Sheridan trans.) (New York: Norton, 1977), p. 96.

35 A. Honneth, *The Struggle for Recognition: Moral Grammar of Social Conflicts* (Joel Anderson trans.) (Cambridge: Polity Press, 1995), p. 74. For further discussion about recognition theory see also the work of Charles Taylor and Nancy Fraser, particularly Taylor's *Multiculturalism: Examining the Politics of Recognition* (Princeton: Princeton University Press, 1994) and Fraser's' Rethinking Recognition', *New Left Review*, 3 (2000), no pagination, http://newleftreview.org/II/3/nancy-fraser-rethinking-recognition, accessed February 2011.

36 This development can be explored in the work of James Gibson, Bruno Latour, Alva Noë and Francisco Varela among others. In particular see J. Gibson, *The Senses Considered as Perceptual Systems* (London: George Allen, 1968); J. Gibson 'The Theory of Affordances', in R. Shaw and J. Bransford (eds), *Perceiving, Acting, and Knowing: Toward an Ecological Psychology* (Hillsdale, New Jersey: Lawrence Erlbaum, 1977), pp. 67–88; A. Noë, *Out of Our Heads: Why You Are Not Your Brain, and Other Lessons from the Biology of Consciousness* (New York: Hill and Wang, 2010); A. Noë, *Action in Perception (Representation and Mind)* (Cambridge, Massachusetts: MIT, 2006).

37 A. Clark, *Supersizing the Mind: Embodiment, Action, and Cognitive Extension* (Oxford: Oxford University Press, 2008), p. xxviii.

14 ✧ A theory of the paint-by-number surface

Stephen Knott

The paint-by-number kit: an accessible art form

The paint-by-number kit is an all in one invitation to art, containing a paintbrush, paints and an outlined image on canvas board with numbered segments matching the colours in the paint pots (Plate 30). There are also instructions that explain how to complete the kit, giving advice on mixing and thinning paints. These kits existed in nascent form in the early twentieth century,[1] but their popularity peaked in 1950s USA when they were sold as an artistic, pleasant and productive use of leisure time.

The story of increasing access to art supplies stretches back to the introduction of drawing manuals in the seventeenth century that persuaded gentlemen and ladies to pursue the 'noble' arts.[2] In the nineteenth century the formation and expansion of art supplies companies like Winsor & Newton, Reeves and Sons and LeFranc et Cie made it increasingly straightforward to buy equipment and materials; their vast catalogues targeted students, commercial artists and a growing number of amateur hobbyists.[3] However, the proliferation of paint-by-number kits represents the apogee of accessibility to artistic production with one advert in 1953 claiming paint-by-number had brought art to 13,475,000 families.[4] The materials for making paintings had long been available separately, but to have them all in one set was unprecedented.

The art supply firm Palmer Paint first developed paint-by-number kits commercially, under the brand name 'Craft Master', in the early 1950s. Although the company was unable to sustain the initial marketing success of the kit, paint-by-number had become firmly entrenched in what social historian Karol Ann Marling described as the 'hyphenated culture' of 1950s America, a reference to the ubiquity of the prefix 'do-it-yourself' in the growth of useful and morally sound leisure activities for postwar suburban families.[5] Firms, celebrities and even the American president Dwight

Eisenhower endorsed this 'can-do' attitude central to the idea of painting your own art.

Existing books by Marling and Dan Robbins have described paint-by-number's social impact and commercial development and many articles since the 1990s have responded to the emerging interest in collecting canvases.[6] This essay, situated within a volume that seeks to define the material surface as a site for social, political and theoretical tensions, has a different priority, contesting the marginalization and presumed superficiality of the paint-by-number surface. Close attention to the material interventions of a diverse array of amateur painters raises the question of the location and meaning of the surface within this medium and how this relates to the role of the surface in wider artistic practice.

The surface intervention of paint-by-number practitioners needs to be situated within the context of modern painting that had acknowledged the potential of the surface to constitute the whole content of a painting. As Nabis artist Maurice Denis proclaimed in 1890: 'A picture – before being a battle horse, a nude woman, or some anecdote – is essentially a plain surface covered with colours assembled in a certain order.'[7] If modern painting was all about surface management, with Denis and fellow artist Paul Gauguin instigating this move towards greater abstraction, what constituted depth in this new aesthetic context? For Denis it certainly was no longer located in conventional notions of technique manifest on the surface, but rather in the elusive quality of artistic expression. Although Denis's words were unlikely to be well known in small-town America, paint-by-number painters were essentially engaged in the same act of material surface intervention.

Reception: Greenberg and kitsch

At its height paint-by-number aroused widespread press attention. Local journalists celebrated the practice as an example of the 'craving for creative self-expression',[8] providing a springboard toward careers in art education for many.[9]

As one might expect, figures of the artistic establishment were much more critical of this dilution of aesthetic standards. In the Midwest, Elizabeth Moeller Geiken, director of the Davenport Municipal Art Gallery, condemned the medium: 'Those numbered paintings evade artistic development completely … A person might as well stamp a pattern on a canvas and call it their own piece of work.'[10]

Geiken's critique echoes the dominant aesthetic critique of the day, underpinned by the ideas of Clement Greenberg. The kits conform to Greenberg's description of kitsch, which 'uses as its raw material the debased and academicised simulacra of genuine culture'.[11] For Greenberg kitsch lacked autonomy, and paint-by-number painters seemingly share this attribute as they heavily depended on other authors. By contrast avant-garde art is replete with its necessary difficulty, self-critical attention to material used, and awareness of the art historical context of production.

Greenberg's division between avant-garde and kitsch follows Kantian principles that defend the autonomy of the artwork rather than its ability to persuade, cajole or stimulate. Paint-by-number paintings of idyllic pastoral scenes and cutesy pets would be placed firmly in the latter category: wallpaper decoration, something to put above the mantelpiece. This accusation that paint-by-number represented little more than decorative wallpaper echoed criticism of Jackson Pollock's 'all-over' paintings. But Greenberg defended Pollock's work. For him, Pollock was able 'to wrench from the decorative the flat, all-over surface and use it in the service of high art', according to art historian Elissa Auther.[12] The elevation of Pollock's painted surface depended on Greenberg's incorporation of his work into the teleological history of the artistic avant-garde and in locating aesthetic depth. Paint-by-number lacked this legitimizing critical apparatus, its surfaces superficial compared to those of Pollock.

There are clear, substantive differences between Pollock's compositions and a paint-by-number of a sad clown, but according to Auther there is an underlying anxiety among art critics concerning the closeness of abstract compositions to wallpaper.[13] As paint-by-number does not pretend to be fine art, an examination of its surface layers facilitates a discussion that can move between material and metaphorical conceptions of surface and depth. As a pared down, exoskeleton of painting, paint-by-number unavoidably accentuates the moments where surface is teased apart from depth, materially and metaphorically, usefully exposing the role of the surface in modern art history.

The centrality of the parergon

To facilitate a sympathetic aesthetic analysis of paint-by-number works, theoretical models which foreground the importance of the supplemental are helpful, in particular Jacques Derrida's critique of Kant's idea of the parergon. Kant described superficial aspects of art as parerga – the colonnades, draperies and frames that surround the central depth of a work. For Kant parerga should be invisible at best, assisting the appreciation of work; at worst they despoil beauty or constitute a work's entire content thus emptying it of depth.[14]

Derrida argues that by naming parerga, an entrenched position between what is intrinsic and that which is extrinsic to the work of art is established. For Derrida the construction of the supplementary is far more fluid:

> It is not because [parerga] are detached but on the contrary because they are more difficult to detach and above all because without them, without their quasi-detachment, the lack on the inside of the work would appear.[15]

Derrida questions the presumed alterity of Kant's concept of the parergon by claiming that it is only ever 'quasi-detached', defining that which is intrinsic to the work of art. Crucially, this more fluid conceptualization of the parergon is not fixed to a material entity: the frame might fulfil the role of the parergon in the traditional gallery, but what happens when the frame becomes the work of art or when a whole room becomes a work of art, as is the case with Duncan Phillips's installation of Mark Rothko paintings within a 'chapel-like' space.[16] A reading of Derrida encourages us to envisage the relationship between what is supplemental and what is intrinsic as a dynamic relationship, a mobile linguistic boundary-setting device. Definitions of essence rely on that which is supplemental and any material can *stand in* to perform this role of marking the quasi-distinction.

Derrida's work is useful in problematizing definitions of surface and depth and helps emphasize the importance of paint-by-number as the 'other' that shaped people's preconceptions of what true art was in the 1950s, a point that William Bird makes in his catalogue.[17] Derrida gives a model that encourages a flexible approach to where one might locate depth and surface, yet relying on Derrida alone is problematic because his analysis is mainly confined to the linguistic terrain and a more material approach to the paint-by-number surface is needed.

Scholarly attention has been bestowed on the material surface as a site of social, cultural, political and aesthetic meaning.[18] A criticism of some of these studies, such as Grant McCracken's essay on patina and Celeste Olalquiaga's poetic evocation of dust, is that they situate the surface as a static category judged in one temporal instant rather than as a mobile and ongoing site of meaning. This volume challenges such rigid definitions. Perception of the narratives and aesthetic depth of the paint-by-number surface demands both an archaeological-geological attention to the different layers on the canvas, and an awareness of different temporal contexts that paint-by-number paintings might pass through. These sensitivities are needed in order to speculate on the location of artistic depth in the paint-by-number surface.

Where's the art?

Geiken claimed above that paint-by-number kits were mechanical, the equivalent to stamping a design onto a canvas.[19] Many other commentators at the time and since have shared her presumption that paint-by-number paintings epitomize the ultimate mechanization of art, 'when thousands of people, many of them adults, are willing to be regimented into brushing paint on a jig-saw miscellany of dictated shapes … and all by rote'.[20] Critics feared the reduction of fine art to an anonymous mass market triviality. However, closer attention to production processes of making paint-by-number kits and their completion reveals the opposite to be true.

Although Craft Master was churning out 75,000 kits a day in the height of the paint-by-number boom in 1954,[21] production processes of paint-by-number relied heavily on hand labour. This was particularly prevalent in the early development of the paint-by-number kit, with Dan Robbins and Max Klein working overtime to work out colour combinations and invent paint dispensing machines, whilst making creative use of available materials (the first pots for paint-by-number kits were made from separated plastic pill capsules).[22] Both in the beginning and throughout the paint-by-number craze in the 1950s, teams of artists were employed to develop compositions, which involved numerous stages (Plate 31). Firstly the artist painted a picture, usually from a photograph. Then the prominent outlines of the image were marked on a transparency laid over the original and numbered. Finally, there was the arduous process of quality control: the intensive study of the painting to make sure the composition 'wouldn't become a helpless mess in the hands of the non-professional', which demanded the skill of managing the simultaneous contrast of colours.[23] Paint-by-number kits were not produced by processes of smooth manufacture; the deconstruction and reconstruction of a painted image had to be done by hand.

The assertion that paint-by-number was akin to a mechanical stamp seems to conform with craft theorist David Pye's assessment of the 'workmanship of certainty', which Pye associates with the exactitude of mechanical production, the lines of the canvas fulfilling the same role Pye reserved for jigs and automated technical devices that ensured a predictable outcome at the end of the manufacturing process. Yet the means by which untrained artists filled in the blanks was far less mechanical than critics like Geiken presumed, and naturally bore the traces of individual practitioners each with their own method of holding the brush, mixing the colours and applying paint. Completing a paint-by-number painting was thus far closer to Pye's definition of the 'workmanship of risk' layered on top of a suggested outline, a jig that is weak in its attempt to impose certainty.[24]

The guidance provided by paint-by-number kits was hardly as dictatorial as critics insinuated: the instructions merely suggested a process rather than requiring its users to adopt them as gospel. Grace Lees-Maffei, in her historiography of manuals and how-to books explains that 'advice has not always been taken seriously', a conclusion particularly relevant to paint-by-number painters who did not always follow the rules of the kit judiciously.[25] Some practitioners might have felt the tyranny of the lines, but many, including Kathryn Bergstrom revelled in the freedom from the rules:

> I didn't have to do it that way just because 'they' said so! What a revelation for a 9-year-old kid in the 50s! I would take the 'leftover' paint from the PBN kits and try my hand at my original compositions![26]

It was art after all. Anything was permitted.

There were several self-conscious means of marking individual interaction with the paint-by-number. Practitioners could mix colours according to individual preference, they could employ an impressionistic technique, correct mistakes in their own idiosyncratic manner and, of course, they could paint over the lines.

Different strategies of approaching the line could be a determining factor in locating artistic depth. Blurring the line, introducing the subtlety of shade and more complex mixing of colours, or going over it with a completely new composition all conform to conventional expectations of greater artistic skill and originality.[27] These modifications are where 'the real art began', according to Bird, 'the moment the hobbyist ignored outlines to blend adjacent colours, added or dropped detail or elaborated upon a theme',[28] and were a source of pride for those who were successful in creating a 'softer look' where lines met.[29]

These material skills, conventionally thought of as higher up the scale of artistic talent, were specifically required in Craft Master's 'Personalised Portrait' paint-by-number kits that allowed amateurs to paint their own portraits after Craft Master artists had created a kit for them from photos sent in. The advertising material for this product emphasized the importance of going over the lines:

> In order to get the soft and delicate gradation so characteristic of skin hues and texture, colours must blend smoothly with each other. Areas which require bleeding are indicated by dotted lines.[30]

Unlike 'amateur-proof' landscapes that were easily reduced to plain blocks of single colour, the portrait paint-by-numbers required extra advice in order to avoid the unattractive over-pixilation of human faces. The introduction of the dotted line to symbolize where bleeding was required demonstrates the fluidity and permeability between qualifications of surface and depth that are *materially* enacted. The ambiguity of an indistinct and vague instruction naturally encourages improvization and more complex consideration of how to achieve a painterly effect with the skills one has available.

The process of filling in blocks of colour that invites comparisons to technical reproduction is only distinguished from skills associated with the more complex, risky and artistic processes of blending and shading through the kit's introduction of the permeable dotted line. The type and quality of material intervention at the dotted line where surfaces blend could be the moment where one is able to separate depth from superficiality: the success in capturing skin tone could be an example of the former. However, pinpointing the exact location where surface becomes depth is problematic. Instead Gilles Deleuze and Félix Guattari's principle of the 'mixture' between smooth and striated space can be applied to paint-by-number: the perpetual cross-fertilization between two abstract concepts either side of a dichotomous division they describe is adequate to the mutual and fluid dependence between surface and depth in the paint-by-number.[31]

Novel and idiosyncratic response to the 'rules' of paint-by-number kits might be an obvious place to identify artistic depth. Nevertheless, these overtly subversive strategies of material surface intervention should not mask the likelihood of risk, subjective craftsmanship and surface depth being evident in all paint-by-number paintings. The traces of individual artistic labour on canvases pregnant with narrative distance paint-by-number canvases from Walter Benjamin's qualification of the arts of technological reproduction. Paint-by-number canvases cannot be viewed as an ironed out flat surface. Plate 32 shows two completed paint-by-number paintings of the same 1969 Craft Master kit *Old Sadface*. There is no attempt in these paintings to blatantly contravene the rules and the painters stay within the lines. Yet this game of spot the difference reveals countless features that distinguish the images: the clown on the left has a lighter hue; there is greater coverage of yellow paint at the top of the t-shirt; a blue line is used to outline the tent pole behind the portrait; and there are many differences between the two faces, notably the complete filling in of the right eye of the clown in the painting to the left which gives the face a more pallid, anaemic character.

The greatest difference is that the image to the right has been authored 'Lynn', a projection of authorship that attempts to add individuality to this commercially produced kit. This authorial mark mirrors a pervasive tradition in modern art to mark authorship, what Baudelaire called the 'poncif', which Ulrich Lehmann recently examined as fulfilling the dual role of designating subjective intervention and functioning as a calling card to advertise an artist's market presence.[32] Lynn's signature does not possess this commercial function, but demonstrates the pride and the desire to author the fruits of one's labour.

The uniqueness of each clown might encourage a capitulation to the idea that every paint-by-number has artistic depth on account of its individuality. Yet limitations to this authorship – the pre-packaged materials and the suggested image on the outline underneath – problematize such conclusions. Instead, the aesthetic interest in paint-by-number derives from the symbiotic relationship between the authored layer and all the material layers underneath, openly demonstrating how painting concerns surface organization and dependence on other labourers.

Surface and depth – a co-dependent relationship

Lynn's signature in Plate 32 shares formal properties with Marcel Duchamp's scrawled signature 'R Mutt' on the base of his unassisted readymade *Fountain* (1917). Comparisons between a painting produced from a kit and a hyper self-conscious avant-garde critique of what constitutes a work of art and artistic labour seem far-fetched. However, they both illuminate a dependence on the labour of other authors and highlight how the material labour of the artist is reduced to finishing off, or in other words, the application of the final layer. John Roberts's description of the impact of Duchamp's readymade can be applied to the paint-by-number canvas:

> The spectator sees – simultaneously – an absence of palpable artistic labour, the presence of the palpable labour of others, and the presence of immaterial or intellectual labour (the reflection, that as a spectator I am being called on to recognize the object in front of me as a work of art).[33]

Lynn's painting was not self-consciously and provocatively offered up as a work of art for aesthetic reception in the same way as Duchamp's readymade: the sad clown probably did not budge from the domestic context until it was sold on ebay. Yet her work evokes the 'palpable' presence of the labour of others. This is at the centre of Roberts's reassessment of the Duchampian readymade as exposing the essential importance of the immaterial labour of the artist. Duchamp's nomination of a mass-produced object as a work of art and the intellectual dexterity and wit to present it as such constitutes the development and elaboration of skill over and above the basis of 'general social technique'.[34]

Like the readymade urinal the paint-by-number painting evokes the labour of multiple authors involved in the production of the art object. The presence of other authors is arguably more palpable in the paint-by-number due to the extent of image fragmentation: the work of artists like Robbins is clearly evident in the final painting, and paint-by-number canvases rarely reached the level of smooth perfection promised by the photographic image on the front of the box and could not entirely conceal the outlined image beneath. More anonymous elements of production – such as those who mixed the paints – are also more obviously exposed because of the kit's limited colour range. The paint-by-number practitioner's dependence on the inexpensive materials in the kit – that would seldom be found in any self-respecting artist's studio – accentuates the centrality of these material aspects of making paintings. Conventionally, trained artists seek to efface the inexpert, raw and even childish material expressions seen in paint-by-number painters' canvases through a more subtle and controlled relationship with the materials they use.

The flourishes, the negotiation of the image outline and the signature constitute the final layer of the paint-by-number, the topsoil of this geology of painting. For the paint-by-number painter 'finishing off' might just involve filling out the readymade pattern. Others might venture further, into the realm of immaterial labour, critically responding to the paint-by-number canvas by going over the line, using different colours, or through a signature that is materially enacted. Given the progression of general social technique represented here by the amateur skill of painting a paint-by-number, the artist is compelled to re-skill in other less materially driven realms of practice. Their finishing off has to be of a greater calibre in the order of intellectual labour.

These advanced skills of finishing off, or the production of the immaterial surface, are shown admirably by Andy Warhol's appropriation of the paint-by-number form in the *DIY Series* of 1962. Warhol was famous for just dealing with the superficial surface and his works can be read as a refutation of the links between the artist and idiosyncratic material skill in favour of inserting the artist into the culture of mass production.[35] Heiner Bastian analyses the *DIY Series* as the logical end point of painting; the skills of the painter are no longer valued in an era where anyone can fill in a paint-by-number.[36] Bastian's observation seems convincing at first, but as I have shown above, even when all the readymades and materials needed to make a painting are handed over to an inexpert audience, the individual still needs to pick up a brush to bring these fragmented parts together.

Furthermore, in the *DIY Series* Warhol did not passively withdraw his artistic authorship through the replication of a popular artistic medium. The work is replete with examples of his intellectual labour, the finish that marks the difference between these works and the huge quantity of paint-by-numbers being completed in towns across America. For example, Andy Warhol leaves large sections of his paint-by-number canvases blank revelling in this unfinished finish (an artistic strategy famously deployed by the French Impressionists in their provocative submission of sketches – *ébauches* – to the Paris Salon as fully-fledged works) and he demonstrates his wit by putting numbers over completed painted sections, getting colour selection wrong and using both crayon and paint. As well as these self-conscious witticisms, Warhol also uses his skills as a commercial printer to compose his sketches, Patrick Smith explaining how painting the whole canvas would have been 'too much trouble for Warhol'.[37] Warhol's version of the paint-by-number is an ironic depiction of a mass-produced art medium.

Yet, his material surface interventions share similar qualities to those of the individuals I described earlier, who broke the rules of the paint-by-number set through a conceptual strategy that was materially enacted.

Conclusion

The superficiality of the paint-by-number surface has been problematized in this essay. Derrida's critique of the parergon provides the theoretical tools to foreground the importance of a marginal form in construction of predominant aesthetic values. The presumed unimportance of paint-by-number also seems questionable in the light of twentieth-century art that has elevated the decorative surface to the realm of high aesthetics, as seen in the work of Denis and Pollock, and situated the readymade as central to redefinitions of material and immaterial skill in artistic labour, as explored by Duchamp.

The power of the paint-by-number surface lies in its ability to demonstrate the dependent but symbiotic relationship between an inexpert practitioner and the labour of other authors, who designed and disseminated the kit. This process alerts us to the fragmentation and dispersal of agency that mirrors the reliance on readymades and general social technique that is to be found in all art, revealing dynamics that Duchamp explored in the incongruous placement of a urinal in the gallery. In the context of disseminated technique epitomized by the ease with which one can follow the instructions of a paint-by-number kit to produce a painted surface, the artist is forced to become more adept in the realm of immaterial labour and must concentrate more fully on materially demonstrating such skills on the outermost layer.

Notes

1 D. Robbins, *Whatever Happened to Paint-by-numbers: A Humorous Personal Account of What it Took to Make Anyone an 'Artist'* (Delavan, Wisconsin: Possum Hill Press, 1997), p. 24.

2 For example H. Peacham, *The Art of Drawing with the Pen, and Limming in Water Colours ... with the True Manner of Painting upon Glasse, the Order of Making your Furnace, Annealing etc* (London: R. Braddock for W. Jones, 1622).

3 L. Carlyle, *The Artist's Assistant: Oil Painting Instruction Manuals and Handbooks in Britain 1800–1900 with Reference to Selected Eighteenth Century Sources*. (London: Archetype Publications, 2001).

4 Craft Master advertisement in *Life* and *Saturday Evening Post* (1953). 'Paint-by-numbers collection, ca. 1934–1957', Smithsonian National Museum of American History, 1996.3019 (NMAH Acc.).

5 'Hyphenated culture' is the title of the second chapter that analyses the social context

of the 1950s paint-by-number boom. K. A. Marling, *As Seen on TV: Visual Culture in Everyday Life in the 50s* (Cambridge, Massachussets and London: Harvard University Press, 1994).

6 For a social history of paint-by-number see Marling, *As Seen on TV*; Robbins, *Whatever Happened* (1997); and W. Bird, *Paint by Number: The How-to Craze that Swept the Nation* (Princeton: Princeton Architectural Press, 2001). For an article charting the collecting interest in paint-by-numbers since the 1980s see M. Daniels, 'Count on it: who would have ever thought those paint-by-numbers compositions would reach art gallery status?' *Chicago Tribune*, 12 September 1993.

7 M. Denis quoted in G. Groom, *Beyond the Easel* (New Haven and London: Yale University Press, 2001), p. 17.

8 A. Treanor, 'Know numbers? To paint a masterpiece', *Toledo Blade*, 22 March 1953. See also 'Art and the numbers racket: Bergson residents burn midnight oil painting with oils', *Bergson Evening Record*, 15 May 1954.

9 The following contributors to the Smithsonian's 'Post-a-reminiscence' website – set up to accompany William Bird's 2001–2 show *By the Numbers: Accounting for Taste in the 1950s* – explain how completing paint-by-number kits has ultimately led to professional careers in art and art education: http://americanhistory.si.edu/paint/reminiscence.html, accessed 24 November 2012. Marsha Rogers (29 June 2001), Kathryn L. Bergstrom (10 April 2001), Julie Hulvey (5 May 2001), Reatha Wilkins (19 February 2002). Bird also defends paint-by-number paintings as valuable in terms of artistic education. Bird, *Paint by Number*, p. 17.

10 'Art world battles over numbers racket', *Sunday Democrat and Times*, 28 March 1954.

11 C. Greenberg, 'Avant-garde and kitsch' (1939), in *The Collected Essays and Criticism* (J. O'Brian ed.) (Chicago and London: University of Chicago Press, 1986), p. 17.

12 E. Auther, 'Wallpaper, the decorative, and contemporary installation art' in M. Buszek (ed.), *Extra/ordinary: Craft in Contemporary Art* (Durham, North Carolina and London: Duke University Press, 2011), p. 121.

13 *Ibid.*, p. 119.

14 E. Kant, *Critique of Judgement* (J. Meredith trans.) (Oxford: Oxford University Press, 2007), p. 57.

15 J. Derrida, *The Truth in Painting* (G. Bennington and I. McLeod trans.) (Chicago and London: University of Chicago Press, 1987), p. 59.

16 Explanation of the 'Rothko Room' on the website of the Phillips Collection, Washington D.C., www.phillipscollection.org/collection/rothko/index.aspx, accessed 24 November 2012.

17 Bird, *Paint by Number*, p. 17.

18 G. McCracken, 'Ever dearer in our thoughts', in *Culture and Consumption: New Approaches to the Symbolic Character of Consumer Goods and Activities* (Bloomington and Indianapolis : Indiana University Press, 1988); C. Olalquiaga, *Artificial Kingdom: On the Kitsch Experience* (Minneapolis: University of Minnesota Press, 2002); A. Gell, *Art and Agency: An Anthropological Theory* (Oxford: Clarendon, 1998).

19 'Art world battles over numbers racket', *Sunday Democrat and Times*, 28 March 1954.

20 Criticism from author of article for *American Artist*, quoted in Bird, 'Every man a Rembrandt', in *By the Numbers*, http://americanhistory.si.edu/paint/rembrandt.html, accessed 24 November 2012.

21 This figure is from an industrial film explaining the process of producing a paint-

by-number kit. National Association of Manufacturers, *Industry on Parade* (18 May 1954), Smithsonian National Museum of American History.

22 Robbins, *Whatever Happened*, pp. 35–46.

23 Quote from National Association of Manufacturers, *Industry on Parade* (18 May 1954); an explanation of the process of making a paint-by-number is given by Dan Robbins in Robbins, *Whatever Happened*, p. 49.

24 D. Pye, 'The nature and art of workmanship', in G. Adamson (ed.), *The Craft Reader* (London: Berg, 2010), p. 342.

25 G. Lees-Maffei, 'Introduction. Studying advice: historiography, methodology, commentary, bibliography', special issue of *Journal of Design History*, 16:1 (2003), 6.

26 Kathryn L. Bergstrom, 'Post-a-reminiscence' (10 April 2001).

27 Reatha Wilkins was so despondent at the equine subject of her paint-by-number Christmas gift that she rejected the outline and created her own painting in 'horse colours'. Reatha Wilkins, 'Post-a-reminiscence' (19 February 2002).

28 Bird, *Paint by Number*, p. 17.

29 Teri Keith, 'Post-a-reminiscence' (18 May 2001).

30 Craft Master's Personalised Portrait paint-by-number kit (1952–53). Mail order form advertisements, *Toy Trains and Hobby Craft* (1953) and *The Toy Yearbook* (1952–53).

31 G. Deleuze and F. Guattari, *A Thousand Plateaus: Capitalism and Schizophrenia* (B. Massumi trans.) (Minneapolis: University of Minnesota Press, 1987), pp. 474–475.

32 U. Lehmann, 'The trademark Tracey Emin', in M. Merck and C. Townsend (eds), *The Art of Tracey Emin* (London: Thames and Hudson, 2002), p. 60.

33 J. Roberts, *The Intangibilities of Form* (London: Verso, 2007), pp. 24–25.

34 *Ibid.*, p. 54.

35 A. Warhol, *The Philosophy of Andy Warhol from A to B and Back Again* (London: Penguin, 2007), p. 101.

36 H. Bastian (ed.), *Andy Warhol: Retrospective* (London: Tate Publishing, 2001), p. 24.

37 P. Smith, *Warhol: Conversations About the Artist* (London: UMI Research Press, 1988), p. 60.

15 ✧ The materiality of print: issues of artifice and integrity relating to the printed ceramic surface

Steve Brown

As a printmaker within the applied arts, I have a particularly intimate relationship with surfaces. Rather than printing 'pictures' on flat planes, my practice is concerned with images that have a direct dialogue with form, defining or determining objects through the physicality of a three-dimensional printed surface.

In 1984 I was apprenticed to a screenprinting company working predominantly with textiles. Over the next fourteen years I progressed from hand printing to large-scale production, finally moving into the sampling of new work, product testing, and interpreting clients' needs. A major part of my pre-production work involved the task of transposing a paper-based image into printed textiles that would eventually become a garment. Looking back, this commercial period of my career was the time when I first began to consider the relationship between print and surface in terms of materials and form. Eventually I left industry to pursue my own creative path, undertaking a series of degrees at the Royal College of Art leading to a PhD titled 'The Physicality of Print'. Over the course of these studies my driving passion was to understand the material potential of printmaking within media such as ceramics and glass.

The use of ceramic materials in combination with conventional printing 'press' methods was developed around 250 years ago. The resulting technique allowed pigments to be 'transferred' from a paper carrier onto ceramic objects. This technique meant that a printed image could be fully fused into the heavy glaze coating of a form, permanently integrating the image with the surface. While other print substrates, such as paper, are fragile and ephemeral hosts for printed information, the ceramic surface is extremely durable. However, such a printed surface may well lack the associations with quality that hand-painted ceramic decoration provides. Mass-production in this sector led to a quantitative explosion, and over time a derogatory view of this area of print became well established.

Negative perceptions of the surface relate to the ideas of the façade, trickery, or deception – surface effects considered as artifice. In the case of a printed ceramic, there is a further problem: before they are fused together the imagery and the form exist in separation, with the print contrived to fit the object. This issue of division is one of the main factors that lead to the pejorative attitude to printed surfaces, as the two primary elements of the design are only 'artificially' united. There is another issue of division, too, which is inherent to most print processes and which exacerbates this physical separation: the division of the process into tasks. As the print critic and writer Pat Gilmour explains:

> An artist's print is not created in the same way as a unique painting, by making direct marks on paper or canvas, but at one remove, by first fashioning what an engineer would call a 'tool' and then using it to produce a series of identical impressions ... The resulting work of art is not the 'tool' described, which is simply a means to an end, but the image such a device is able to transfer repeatedly to [the substrate].[1]

The original work is separated from the surface by many removes, long before the print is finally effected. As a result the process is not spontaneous (and might well come to seem insincere). In my own experience, I have found that screenprinters move their original imagery through a series of transformative phases: to film, then onto a screen, and finally onto the substrate. To work in this way, at a number of 'removes' from the final result, requires the practitioner to employ sophisticated predictive interpretation. These degrees of separation in printmaking are compounded further when we move to production within the ceramic industry. Here the printed image may undergo a further transposition from the original materials into ones suited for ceramic processes, as well as a transformation from two dimensions to three.

The Staffordshire ceramic manufacturers Royal Crown Derby specialize in this area, producing printed ceramic objects to the highest quality. The division of labour there is extensive. Designers originate the concept drawing from which two paths develop. On the one hand, the form is produced by model makers and mould makers, then brought to the casting shop, glaze lab and finally the kiln room. In parallel to these activities, the surface imagery is developed by separators (who interpret the concept designs into imagery that fits the form), colour mixers, print samplers and lithographers (who apply the transfers). The thorough division of labour at Derby results in a strong separation between the various aspects of form and surface, which in turn requires constant checking: the printed imagery is continually offered up to the form to ensure that it fits exactly, creating what might be called an illusion of integrity. For many companies, this time-consuming level of sophistication is simply not expedient, and cheaper decorative objects often suffer from a mismatch of object and imagery.

It is not just low quality that can spoil our appreciation of a printed ceramic object; a badly conceived relationship between image and form is the simplest way to undermine the integrity between the two. One object that illustrates this is a simple print decorated plate from the Victoria and Albert Museum's ceramic collection, printed with a pattern called 'The Sheep and the Crow' by Sadler and Green for Wedgwood.[2] In this case the relationship between the image and form has been determined by economy rather than design. In the eighteenth century a copper plate would have to be engraved in order for a print to be made – a long and costly process. The print that has been applied to this particular plate was originally designed to decorate a square tile; in an attempt to squeeze more value from the design it has been re-used on a circular plate, with various motifs added to fill the gaps. The result is that the two components, print and object, are at odds with each other.

Another issue in ceramic surface design is the changing perception of decoration, which has at times been hidden within the shadow of modernist formal doctrine and the rejection of ornament. The modernists attached a moral value to the notion of 'undressed structure', which they understood as honest form. A major crime was imitation. Victorian reformers like John Ruskin decried 'surface deceits', the inappropriate 'painting of surfaces to represent some other material than that of which they actually consist'.[3] The writer Leonard Whiter similarly praised the magnificent raised solid gold surfaces produced by hand at Sèvres, describing them as 'not so much effect as fact', and criticized cheap Staffordshire copies in gilt paste.[4] If print is not used for its own unique strengths it becomes simply a second rate reproduction.

Although there are several technologies that can be used to print a ceramic surface – relief, intaglio and lithography, all are saddled with the problem of pressure. Though it is common to speak of printing as something that happens on a 'press' (which aptly describes the action of applying extraordinary pressure onto the print matrix to create an impression), this is not an appropriate way of directly printing onto fragile ceramic works. Instead, an indirect transfer system must be used. Because one cannot apply large amounts of pressure it is not possible to lay down heavy deposits of ink. Ceramic printmakers overcame this problem for many years by using extremely rich pigments such as the ubiquitous cobalt blue, which only requires a light deposit to produce a satisfactory image.

Another technique, developed in the mid-eighteenth century specifically for working with ceramic forms, employs a flexible gelatine bat which could be used to print a curved ceramic surface.[5] This method was mechanized by

Guy Murray in the mid-1950s employing a pad that was cast as a parabola. Using this approach, which came to be known as 'direct printing', the pad could pick up the print from a flat inked plate and deliver it immediately onto the surface of a curved object.

It was, however, the screenprinting process (a development from the open stencil technique) that revolutionized the relationship between ceramic materials and the printed surface. The direct relationship between pigment and substrate that this process offers is not compromised by excessive pressure. Ceramic ink is more gently 'laid' onto the substrate's surface, controlled by the stencil mask, with only a thin mesh support to obstruct the ink's flow. Screenprinting allowed the ceramic industry to fully exploit the abundance of ingredients on the glaze lab shelf. Heavy printed layers gave rise to a more expansive range of colours and rich surfaces. Alternatively, thick reactive glaze materials could be laid onto the ceramic surface. Rather than reproducing decorative designs or pictures, these material deposits would change dramatically during firing, subtly shifting and flowing like watercolours. Each printed surface within a large batch was slightly different, departing from the mass-production aesthetics usually associated with printing. Screen techniques allowed prints to be better integrated with the ceramic surface, and raised the perceived value within the industry. The problem still remains, however, that the printed image and ceramic form remain divided, combined only at the very last stage of production.

Studio practice is not defined by the same means and aims as large-scale commercial production. Individuals working with ceramics and print have at times looked to overcome the physical dichotomy between the two, by using strategies such as printing directly onto clay and then manipulating these panels into simple forms. It is in the studio that we can break away from the conventional approaches necessary for industrial reproduction. One of my main practical objectives is to find new ways to fully integrate the printed image and ceramic form. In an early body of work, I dealt with this challenge by screenprinting images directly onto flat plaster moulds. The images could be intuitively collaged, as could the final construction of the moulds. This approach integrated the printed image with the ceramic surface at the making stage, and offered some versatility, allowing creative choices that affected image and form together. With the idea of 'in-mould-decoration' established, I became interested in taking the relationship between clay and imagery further, to a point where they could be manipulated together. Ultimately, I aimed to be able to 'sculpt' with my prints.

By printing onto plaster moulds, I replaced the conventional transfer paper with a substrate that could be constructed into angular architectural forms. My next step, based on my experience in the textile industry, was to print ceramic pigments onto a flexible fabric. This could in turn be constructed into far more complex moulding structures by simply stitching the panels together. When pushed into this envelope the amorphous clay body becomes formally determined by the textile, creating a hybrid surface that is an integrated composite of clay, ink and physical barrier, all of which can be stretched, manipulated and formed as a united surface/form. Any transformation is effected upon both aspects simultaneously: formal plasticity integrates with surface elasticity. Each adjustment of mass, volume and form is inflected by the sophisticated skin that surrounds the clay body, as it is drawn out, twisted and distended. This approach records the physical event of making, a spontaneity that is denied by the traditional 'removes' of print production.

When finished and dry, the work is kiln-fired, producing a final alchemical transposition between the clay and ceramic states. The fabric envelope is incinerated, leaving only the vestigial layer of imagery fused into the surface. My aim is to create an undivided surface/form composite, a whole that is *übersichtlich* – visually easy to grasp. Through this new process, I have also inverted the traditional hierarchy that favours ceramic form over surface. Often deemed unnecessary and interchangeable, a surface print can be thought of as supplementary to form. A ceramic object can exist without its printed surface, but not vice versa. My approach begins instead with a fully resolved, self-supporting printed image. This print has the potential to exist as a self-standing artwork, rather than as an adjunct to form. Only once the print is finished do I begin to make it into an envelope matrix, which will bring about a new form. While this surface image dominates and predetermines the underlying form it is not developed in isolation but rather with the end object in mind.

One of the final works that resulted from this experimentation took as its inspiration the mandrake, a plant whose root was traditionally attributed mythical properties (Plate 33). Like other tubers, such as ginger, a mandrake root gives the impression of being a hybrid form with both animal and vegetable qualities. It can be thought of as existing on the threshold between two organic states. Its growing position, both above and below the surface of the earth, further suggests a liminal state. I drew upon these ideas relating to transposition when making the mandrake series. In one of these final works, the two-dimensional printed textile image (that is, the determining matrix) is displayed in combination with the three-dimensional, printed ceramic object that results from it. The work occupies two states at once, emerging through the surface of the print to become form. In designing this piece, I made decisions relating to the 2D image/skin based upon predicting the effect it would have upon the 3D form. I was not simply designing a 'picture'; every choice would have a structural impact upon the final form. The process of formal development focused on two elements: the branch – and finger-like 'fronds', and the spiralling tail of the tuber. The spiral is a good example of the subtle changes that occur through the process; a curling tail emerges from the flat textile spiral as it is invested with clay, 'corkscrewing' outwards as form.

The *Mandrake* series also demonstrates another aspect of the surface that has become important to my work: the seam. This is a formal element determined by the joining of two or more textile components.[6] In my work the seam has become the physical and symbolic site where the division of surfaces is overcome. The seam defines the edge of the textile panel and the printed image and so exerts a major influence on the final form. It is at the seam that I am able to disrupt the planar surface with lumps, tendrils and outcrops of form. It is also at the seam where the two mirror images come together to form a new symmetrical image; the edge becomes the centre. The nature of a seam is to emphasize a juncture, and I amplify this effect through surface imagery. In the *Mandrake* works, the flow of the corkscrewing tail is reinforced through the spine of the seam running along the median line of the body, and this is accentuated through the fluid inflection of the surface imagery emanating from this furrowed, disruptive depression.

Another idea that has occurred to me since I began working with printed fabric envelopes is that these printed surfaces act as a type of garment for the plastic clay inside. This idea inspired my work *Arlecchino the Brave* (Plate 34). I designed a bright, brash surface decoration to include large areas of diamond pattern work, which is all that is needed to associate this abstracted figure with the character from the *commedia dell'arte*: the clumsy fool. In this case it is not just the manipulation of the form that is inflected by the surface imagery, but also the character the work portrays.

This technique can be considered as a low-tech, manual version of digital 3D printing technology, often referred to as Rapid Prototyping. We can think of these processes, too, as a kind of printmaking, though they have leapt away from the printed surface. The previous distinction between ink and substrate is abandoned, and the boundaries between printing and casting are blurred. The new areas of printmaking I have explored, in which the process is brought into three dimensions, operate according to a fluid conception of 'surface'. My surfaces are in constant flux, both carriers of visual information and envelopes that determine formal limits. Once my pieces are fired, one surface disappears and another is created, as it merges with form as a trace or record of the making event. My research has led me to understand the surface not as a mere canvas, dominated by a representational picture, but as the material and physical domain for expanded printmaking.

Notes

1 P. Gilmour, *The Mechanised Image: An Historical Perspective on Twentieth-century Prints* (Twickenham: CTD, 1978), p. 271.

2 'The Sheep and the Crow' printed by Sadler and Green for Wedgwood. V&A Ceramics Collection, museum number: 414:1135/J-1885 1775.

3 J. Ruskin, *The Complete Works of John Ruskin*, vol. 1, *The Seven Lamps of Architecture* (New York: Thomas Y. Cromwell, 1905), p. 32.

4 L. Whiter, *Spode* (London: Barrie and Jenkins, 1970), p. 181.

5 A. W. Coysh, *Blue Printed Earthenware 1800 –1850* (London: David and Charles, 1972).

6 For an early discussion of the seam as a critical juncture in design, see G. Semper, *Style in the Technical and Tectonic Arts* (Los Angeles: J. Paul Getty Trust, 2004), p. 153.

16 ✧ Substance abuse: the postmodern surface

Glenn Adamson

As one of the curators of a major exhibition on postmodernism, I have often felt like a cartographer trying to draw a completely reflective surface, which repels any attempt to verify its contours.[1] Or like Orson Welles, shooting the final scene of *The Lady from Shanghai*, in which the heroine and the villain of the film are trapped in a 'magic mirror maze' at the funhouse.[2] This climactic scene culminates in random gunfire, which brings the myriad images crashing to the floor. In the same way, postmodernism has always presented onlookers with a dizzying refractory play, which resolves only in its collapse, and only in hindsight. In this brief review, I will trace a somewhat arbitrary route over this heap of fragments, arguing that it was the surfaces of the postmodern era – whether glossy and glamorous, or distressed and dystopic – that constituted its very core.

As the very word suggests, postmodernism is premised on uncertainty – the uncertainty of a phase in history that has been foreclosed without a new one really getting underway. Even in the years of its emergence – the years prior to 1977, when Charles Jencks gave a name to the insurrectionary tendencies then rife in architectural practice – the only certainties that attended postmodernism were framed in negative terms.[3] Away with rationalism, away with narratives of progress, away with objective standards of fitness. All well and good; but what replaced these hard-won modernist principles was a sense of aimless drift, in which the history of each medium became available once again for raiding and quoting. Beyond this, there was only the gesture of negation itself.

Yet this negative energy was the combustive fuel of postmodern practice. The energy of the moment was a sort of raw kinesis, a loose force released through the act of breakage. Accordingly our exhibition began with images of destruction: a chair set afire by Alessandro Mendini (Plate 35); the Pruitt-Igoe development tumbling spectacularly to the ground, embraced by Jencks as an anti-monument to modernism's failure; a grand plaza imagined by its own architect, Arata Isozaki, in a state of future ruination.

Anyone who has been a teenager, or lived with one, will recognize the fundamentally adolescent nature of these gestures. In its initial phase postmodernism was a *cri de coeur*, a simple statement of radical disaffection, rather like the undisciplined shout of a punk rocker – which was, of course, another part of the cultural landscape circa 1977. And yet, like punk (or a sixteen year old), postmodern practice contained strains of comic insight, as well as the possibilities of future sophistication. Within a few short years, postmodern discourse would dominate every area of design, not to mention performance, film and television, music, and fine art.

Orson Welles's cinematic metaphor of the broken mirror – a surface in parts – is helpful in visualizing this transformation. When a blow strikes reflective glass, the immediate result is expansive unpredictability. But the new topography thus created, as dazzling and self-contradictory as it might seem, can indeed be mapped with difficulty, though it will constantly shift according to one's perspective. Similarly, after its initial formation through acts of intemperate violence, postmodernism did become legible. Its internal ruptures became vectors for travel. Fragmentation itself became a technique, as did gluing together, and 'making up', in every sense of that phrase. No longer nihilistic, postmodernism instead became superficial. Across its multiple, conflicted surfaces played a distanced and post-lapsarian look– a state of heightened yet disjunctive self-regard. This fragile postmodern surface, a glittering façade of fragments, became, in turn, the unstable foundation for our own pluralistic moment.

This brings us to the vexed matter of periodization. The years covered in our exhibition were, more or less, 1970 to 1990 – after the technophiliac Pop associated with the counterculture, and before the onset of digital design and the Internet. This positioned our project comfortably after the design era usually referred to as 'mid-century modern'. Yet the 'post' in postmodernism can be understood in at least two ways. It could simply mean 'after', and for a curator that is an attractive prospect. It would mean that postmodernism is a distinctive and separate movement, a genuinely new set of tendencies which cannot be subsumed within an overarching narrative of modernism. But one must also contend with another version of the 'post', which would imply that it is a mere adjunct, like a footnote or epilogue – a postscript. This view, which is (not incidentally) held by many of the practitioners we included in our exhibition, involves diminishing or even denying postmodernism's claims to radical departure, instead relegating it to the status of a late-arriving, mannerist spasm within modernism itself. These two positions may seem to be contradictory, but as often with postmodernism, it's not really necessary to choose between apparent opposites. There is no necessary contradiction between the desire to overthrow modernism's repressive aspects and the desire to sustain its restless inventions. Many practitioners, indeed, can be seen trying to achieve just that, which accounts for the disjunctive nature of their work.

It is with this Janus-faced idea of postmodernism in mind that one can best understand its surfaces in depth. The logic of the façade as the dictating element within a design betrayed a fascination with external appearance, rather than the underlying structure or skeleton emphasized by modernists. This shift to the surface was first propounded in a serious way by Robert Venturi and Denise Scott Brown, whose writings (notably 1972's *Learning from Las Vegas*, written with colleague Steven Izenour and based on a research trip to study that city's neon-strewn strips) put forward the typology of the 'decorated shed' as a model for communicative architecture. As they pointed out, in structural terms a casino is not really that different from a baroque cathedral. Both are big rectangular volumes that would be anonymous without their façades. The buildings are literally front-loaded, announcing their meanings as a human face projects cerebral intent. This was a powerful idea that took an unexpected course over the ensuing decades. The trajectory is perhaps too well known to review here.[4] It suffices to mention certain key architectural statements of the 1980s. Michael Graves's Portland Building was like four billboards joined at the corners, its neo-Art Deco ornament pinned to its sides like butterflies under glass. Philip Johnson's controversial skyscraper for AT&T was lavishly veneered inside and out, a confection of gold leaf, dark wood and pink marble, and its 'Chippendale' top, a split pediment lifted high above the Manhattan streets, was like a corporate logo stamped on the skyline. And then there was John Portman's Bonaventure Hotel in Los Angeles, proclaimed as prototypical postmodern architecture by no less an authority than Marxist theorist Frederic Jameson. The Bonaventure exemplified the 1980s fad for buildings sheathed in mirror-like glass that recapitulated their own volumes, and those of the city around them, in ever-shifting, fragmentary form. (Johnson, with his eye ever on the main chance, had been one of the earliest exponents of this style, pioneering it in such buildings as PPG Place, in Pittsburgh.)

Whether ornamental or reflective, postmodern façade-ism was a corporate style *par excellence*. As Reinhold Martin has recently argued, there was even rhetorical profit to be found in the myriad reflections seen in this style of architecture, which he prefers to read as *de*-flections: 'The tricks with mirrors and other real materials performed by corporate globalization produce the illusion that there is an illusion: the illusion that their materiality is illusory, unreal, dematerialized … This mode of distraction draws us in even as it keeps us out.'[5] Venturi and Scott Brown certainly did not approve of these applications of their ideas, and for good reason. The tidy graphic force of Graves's and Johnson's buildings was diametrically opposed to their call for architectural surfaces made up of contingent parts: what Venturi had earlier called the values of 'complexity and contradiction'.[6]

The annals of architectural theory are littered with debates about the postmodern façade; in that sense, it has been extensively surveyed. But it is worth underlining the point that it was on the *surfaces* of buildings that the two faces of postmodernism, its apostasy and continuity, converged. All the figures mentioned above were saturated in the International Style. Johnson had, in his youth, even curated the groundbreaking shows at the Museum of Modern Art that first introduced that term, and along with Venturi and Scott Brown, and Portman too, he thought of himself as extending modernist practice, not undoing its precepts. Even Graves, the principal avatar of American postmodernism, was doused in modernism and never abandoned some of its humanist ideals.

This convergence of two senses of 'post-ness', and the political and aesthetic tensions that existed between resolved and fragmentary surfaces, were also played out in other disciplines. This is a less well-known story, and one we tried to tell in our project. Over and over again in postmodern material culture, one finds the metaphor of the 'decorated shed' echoed in other forms. Slogan t-shirts, of the kind sported by designer Katherine Hamnett to critical effect, found their corporate equivalent in branded sportswear emblazoned with corporate logos. Such attire rendered its wearer into the bodily equivalent of a building composed of a 'dumb' box with a highly 'articulate' façade. And just as in Venturi and Scott Brown's architecture, there is room for intelligent play in this relation. Wearing a Gap t-shirt certainly marks a certain relation to capitalism, but leaves the exact nature of that relation undefined. It is this indeterminacy that has allowed hip hop culture (for example) to engage in a constant, teasing dance with corporate brands. In hip hop clothing, the logos of companies such as Adidas and Tommy Hilfiger are first appropriated by kids on the street (or more latterly, media-ready performers). These companies then market themselves actively to the hip hop audience, who may well react in gestures of satire or rip-off. This oscillation is matched by hip hop's multiple inhabitation of the 'post'. On the one hand, musicians like Afrika Bambaataa and Grandmaster Flash extended modernist practices, such as Duchampian appropriation (seen particularly in sampling) and the fetishizing of technology. On the other hand, hip hop never involved resistance to the marketplace, as the modernist avant garde did; its representations of sexuality, race and wealth are ironical restatements of existing hierarchies, which unsettle through exaggeration rather than direct challenge. So, as Venturi and Scott Brown might put it, hip hop is both/and, at once a realization of modernism's promise and a turn to the completely new possibilities of a contingent subculture.

Another case of over determination in the postmodern surface can be found in the concept of bricolage. The shattered mirror of postmodernism was depicted in surprisingly literal form by graphic designers, as in April Greiman's 1979 cover design for the west-coast lifestyle publication *Wet: The Magazine of Gourmet Bathing*, realized with the help of her then partner, the photographer Jayme Odgers (Plate 36). At the core of the image is the lipsticked face of a teen idol, surrounded by a whirlwind of jagged fragments. Stray markers of possible identity, such as a Japanese Noh mask and a pair of marbles (not to mention two bar codes for the price of one), are set within an otherwise abstract representation of deconstructive fluidity. All this is quite distinct from modernist practice. Yet Greiman initially studied with Wolfgang Weingart in Switzerland, and was thus exposed to a grid-based technical system (as well as its possible deconstruction). Moreover, whether working in cut-and-paste (as when she and Odgers made the *Wet* cover) or with early digital tools, Greiman subscribed to the venerated modernist principle of treating technique as content, whether it was glued paper or the individual pixels of a rudimentary typeface on a 1985 Apple Mac computer. In general, it is difficult to distinguish postmodern bricolage from the modernist collage that preceded it. On both a technical and conceptual level there is not much difference, arguably, between the collaged works of Pablo Picasso or Hannah Hoch, and the bricolages of Greiman or other postmodernists, such as the painters Robert Rauschenberg or David Salle, or Frank Gehry in architecture. All derive energy from the act of tearing. All benefit from the unexpected associations caused by arbitrary joining, the open suggestiveness of random juxtaposition. And all reflect on the way that life is already occupied by a free play of representations – hence Picasso's witty appropriation of the letters *jou* (game) from the title *Journal*.

So what's the difference? It is again a matter of attitude to the surface. Take for example the postmodern stylist Ray Petri's spread 'The harder they come the better', published in the magazine *The Face* in 1985 with photos by Jamie Morgan. In the most famous image from the shoot, child model Felix Howard is given the supremely unfitting appellation 'killer'. The image is made up of pre-existing parts, as surely as any Synthetic Cubist composition. There is the zoot suit tailoring, signifying gangland cool; a feather and handkerchief denoting upper-class dandyism; a photographic manner suffused with film noir chiaroscuro; and of course, the miscast Howard himself, who is deployed like a living found object (and given the publication context, perhaps as a token of youth culture *per se*, as Carol Tulloch has observed).[7] Picasso used the constituent parts of his collages to reflect on the churning sea of signs that he unexpectedly found at his own breakfast table. Hoch turned a similar formal vocabulary to incisive political critique, dissecting media-propagated ideas about class, gender and officialdom. Petri and Morgan, by contrast, present us with a completely fabricated fiction in which there is no clear social referent. There is no object of critique here, nor can the parts be made to cohere into a higher truth, even about representation itself. There is no insight about the assorted fragments; no realization provided by the unifying field of authorship; no message about the nature of representation. It is rather the fact of fragmentation that is presented to us, in thrillingly distracting rather than cerebral form. In place of ideas, we have style itself, conceived as fissile material.

This approach to the fragmentary surface, in which the joints are left raw and exposed, is one of the key traits binding together postmodern practice. If modernist collage operated according to a method of matching unlike things, creating a new whole, then postmodernist bricolage is all about the mismatch. What lies between the fragments, in the divisions of these surfaces, is sublimated antagonism. Charles Jencks wrote of postmodern architecture as being held together with 'smash joints'.[8] In the realm of aesthetic theory, Derrida insisted that the apparently clean cut that divides one aesthetic form from another, as in the demarcation between a work of art and its frame, is a site of power and force: these cuts bleed.[9] The New York choreographer Karole Armitage, in her 1981 work *Drastic Classicism*, put her dancers onstage with four guitarists and one drummer (Plate 37). To the thrum of an almost painfully loud wall of noise, they cavorted across the stage, striking one pose after another: arabesques drawn from the classical ballet technique in which Armitage had been trained; indicative movements quoting the work of avant garde doyennes Yvonne Rainer and Trisha Brown; and improvised flailing, straight off the punk rock dance floor. Like Greiman, Petri, Gehry, or Salle (who was her partner and frequent collaborator at the time), Armitage made no attempt to patch the transitions in this astonishing pile-up of dance idioms. The violence of the music and the movements in *Drastic Classicism* was forceful, even disturbing, and underpinned a sense of constant formal rupture.

Thus, the nihilistic breakage endemic in early postmodernism was elaborated into a generative technique. For a choreographer like Armitage this was obviously a three-dimensional affair, but postmodern method was more often played out across the surface of objects. The most obvious example is radical design from Italy in the late 1970s and early 1980s. Studio Alchymia, led by Alessandro Mendini, incorporated discrete elements created by different designers into their *Mobile Infinito* ('infinite furniture') series. As if they were covered with a kind of theoretical adhesive, each cabinet and table juxtaposed numerous authors without any explanation of their collaborative intent. At this time Mendini also concocted his Proust chair, another paste-up job: a title taken from literature, a form adapted from eighteenth-century Baroque furniture (albeit swollen to improbable proportions), and decoration swiped from a Pointillist painting by Paul Signac (Plate 38). This surface treatment was achieved through the ingenious means of projecting a slide of the painting across a room at the chair, and daubing paint on to it to match the dots. One can almost hear Mendini saying, 'this is a stick-up'.

Alchymia's repertoire was brought to its logical conclusion in Memphis, which immediately upon its launch in September 1981 achieved totemic status as the ultimate in postmodern design. As Catharine Rossi has shown, the Memphis group operated according to the precept that its leader, Ettore Sottsass, called 'glue culture'.[10] His own furniture, ceramics, metalwork and glass, like the designs by his younger colleagues in the collective, were put together at uncomfortable angles, without transition or nuance. Colour was used to create maximum clash, and was always used in solid blocks or patterns, rather than in modulated tones. Memphis designs are often described as 'toy like', but if that is so, they seem to be made of parts taken from a number of different kits. The influential furniture that Sottsass introduced in the first Memphis collection – like the *Casablanca* sideboard – seemed like decorator's chips tossed into mid-air and glued along their edges. And those edges were hard to miss. The plastic laminates that formed the surfaces of these pieces were printed only on their front face; in depth, they were dark and opaquely resinous. This meant that the joints of the furniture were literally 'black marks', like those assigned by a disapproving teacher. They gave the objects a ticky-tacky, obviously ersatz feel, the diametric opposite of modernist furniture which had been designed (as it were) all the way through. This method of assembly survived a number of shifts over the course of Memphis's few years of operation. In 1982 Sottsass and his colleagues began exploring luxury materials, including marble, often juxtaposing the real material with cheaper stuff painted in a *faux* finish. In the middle part of the decade into the early 1990s, he did a series of furniture designs which nod to both Cubism and classical ruins, but tumble so wildly through space that they bring to mind Walt Whitman's famous phrase, 'I sound my barbaric yawp over the roofs of the world'(Plate 39).[11] Even in these assertively self-expressionist works, however, Sottsass embraced the signs of artificiality, plastering every square inch with applied ornament.

The objects designed by Sottsass and his colleagues in Memphis were quintessentially superficial, props made largely for the purpose of achieving an image in a glossy magazine. As the ceramist and designer Peter Shire (the group's most enthusiastic American participant) has observed:

> Memphis was *of* the media. There was never any problem with colour separations, it always reproduced true, because we were using synthetic

colours in the first place. The priority was to go for the image. The difference was between its existing and not existing.[12]

Postmodernists brought this cosmetic impulse not only to objects but to the body as well, especially in the domain of pop music. As the names alone suggest (Television, Visage, Talking Heads, *The Face*), New Wave music culture was obsessed with identities that were painted-on for the purpose of being broadcast through mass media. The extraordinary make-up worn by performers like Grace Jones was partly a matter of postmodern overstatement, but it also indicated a radical disjunction, another violent cut, this time between the authentic self and the performing subject. (As one Jones lyric had it, 'Fame, all alone with my name, even that don't belong to me.') In the music video for the 1980 song 'Fade to Grey', by the pop group Visage, cosmetics are made explicitly into a metaphor for the deconstructed subject, as the singer Steve Strange is broken into fragments through montage: an eye and a pair of rouged lips, patches in the darkness slowly coalescing into a face. Then, true to the song's title, he fades away into nothing.

One of the great dangers of revisiting such examples from pop culture is that they are likely to prompt nostalgia; an MTV video that may have been provocative or even alarming when new, like *Fade to Grey* or Devo's ultra-artificial, sadomasochistic video for the song 'Whip It', no longer compels us to rethink things. Instead it prompts an affectionate reverie about years gone by. This prompts the thought that it might be worth comparing postmodern design to the less lovable things about the 1980s: for example, the decade's embrace of 'hard' drugs. All the design, art, music and architecture in the world can't quite replicate the shock we still feel when we see Al Pacino in *Scarface*, with his face planted in a pile of white powder; or, in an opposite but perhaps equally appalling cultural cliché, recall Nancy Reagan's nannyish warnings that Americans had better 'Just Say No' or else; or read a novel like Bret Easton Ellis's *American Psycho*, in which the serial killing anti-hero flits around Manhattan, torturing his victims, eating nouvelle cuisine at upscale restaurants, carefully parsing Huey Lewis and the News albums, and snorting high-grade cocaine in the bathroom of a club. As Ellis's novel makes abundantly clear, there is a certain version of the 1980s (an American, urban template) in which morality has fled and surface is all that is left. Characters are described only in terms of their designer clothes, not in terms of their faces or personalities. And drugs were part of this thoroughly superficial, synthetic affect.

Indeed, the distance between the 1960s and the 1980s could not be captured more economically than in the contrast of drugs associated with those decades.[13] On the one hand, the late 1960s counterculture had organics such as marijuana, the peyote cactus and psilocybin ('psychedelic') mushrooms, delivered in more or less natural form, or in the special case of LSD, a synthetic that carried with it a utopian message of liberation and was presented to a user in the form of gaily pictorial tabs. In the 1980s, by contrast, we have, above all, cocaine – a natural product, but one that has been processed into antiseptic featurelessness – and a science fair's worth of synthetics, most of which were experienced by consumers in the form of brightly coloured pills. Hallucinogens and depressants were still in wide use, but stimulants joined the party, and in many cases dictated its flow. Things that grow in the ground yielded to artificial products made in a lab; and drugs that imply a 'deep' experience in a communitarian context were replaced by ones that were thin and sharp, and intensely individualistic, even when taken in a dance club. If 1960s drugs took the edge off, 1980s drugs were all about gaining an edge – often, a competitive one. The very term 'hard' drugs captures this idea of the postmodern 1980s drugs nicely: these were substances that matched the period's driving capitalist ambition. They were not gateways to new experiences, but means of intensifying the sweet, shallow feelings of success you might be having anyway.

So what does the psychological superficiality afforded by certain drugs have to do with postmodern design? We might begin with a striking list of the similarities between cocaine and Memphis – which counted an all-star list among its members, including high-profile contributors from America (Michael Graves), Spain (Javier Mariscal), and Japan (Arata Isozaki and Shiro Kuramata). Somewhat facetiously, then, we could describe Memphis as an international cartel; more seriously, we might simply note that their works made their way to the high-style interiors of London, New York, Paris and Monaco. And that geographical movement, like that of cocaine, was passed on by those in the know. The very possession of this design contraband imparted an intoxicating effect of energy and mastery, recalling David Banash's observation that 'cocaine legitimates the preferred modality of capitalist subjectivity – radical and inviolate individuality'.[14] That line could have been on Ettore Sottsass's business card, and indeed he had long been interested in reflecting on drug use in his work. In the late 1960s, he produced fantastical ceramics for the storage of his pills, writing that he wanted to 'make myself a filling pump where I can fill up forever on 4-star fuel, fill up my veins and set them alight'.[15] In the 1980s he viewed his own work as aimed at a no less intoxicating effect: 'The only thing I know is that Memphis furniture is very intense and that it can only live with very intense people, very highly evolved and self-sufficient people.'[16]

Finally, there is the matter of artificiality itself. In the 1989 film *Drugstore Cowboy*, one character has the lines: 'Most people don't know how they're gonna feel from one moment to the next. But a dope fiend has a pretty good idea. All you gotta do is look at the labels on the little bottles.'[17] This is typical of one kind of postmodern thinking. It is one thing to note that drugs artificially reproduce 'natural' emotional states; it is another thing entirely to conclude that drugs contain within them *all* potential emotional states, and can therefore be considered replacements for emotional life in general. This is the logic of the simulacrum, familiar to readers of postmodern theory, or even to anyone who's seen the postmodern films *Blade Runner* (1982), *The Truman Show* (1998) or *The Matrix* (1999). In all these cases identity is revealed as illusory, a matter of collecting together a given set of pre-existing tropes. This was the philosophy, too, behind such works as Charles Moore's Piazza d'Italia, which quotes the history of classical architecture willy-nilly; or Venturi and Scott Brown's iconic plywood chairs for Knoll, in which particular historical periods, such as Queen Anne, Gothic, Biedemeier, or Art Deco, were applied like so many Post-it notes.

This cynical, quick-hit version of postmodernism has earned many enemies – among them the modernist architectural theorist Bruno Zevi, who raged that 'Whoever decides to abandon the modern movement can choose between Versailles and Las Vegas, between sclerosis and drugs', and the Tudor historian Sir Geoffrey Elton, who dismissed postmodern literary theory in 1990 as 'the intellectual equivalent of crack'.[18] Such charges may seem cheap, but they contain more than a grain of truth. Postmodern design at its height was a particular kind of indulgence. These commodities not only incited desire, but rendered it in something like a pure, self-reflective state. Just like cocaine, once unleashed on the world postmodern design was a genie out of the bottle. It seemed to proliferate according to an economic imperative of its own, with individual middlemen simply the handmaidens to its diffusion. Gradually design in this period (the so-called 'designer decade') was reconceptualized so that it came to refer to *any* reconfiguration of production that took a step ahead of existing frameworks. This could be a matter of creating a new 'designer drug' (like ecstasy or China White), but it could also mean rearranging an office layout to shift expectations about hierarchy; or rethinking the parameters of the body through dress in order to subvert norms of sexuality and propriety, as in avant-garde Japanese fashion or, at the other end of the spectrum, the power dressing seen on a TV show like *Designing Women*; or redesigning experience itself, within a mall or a computer game simulation, or in the case of Disney, perhaps the most widely recognized postmodern design firm of all, across a network of commercial persuasion extending from theme parks to themed stores to theme songs.

We might sum up this new, intangible, borderless idea of what constitutes design with the indelible but much overused 1980s phrase 'thinking outside the box'. Today, design theorists routinely speak about their subject matter in terms of networks, systems, or social structures, rather than concrete objects.

In this respect, the discipline is merely keeping up with the characteristic forms of late capitalism, which tend to approach the immaterial. Advertising imagery, flotsam-like celebrities, junk bonds, electronic music and perhaps even post-conceptual art were all part of this moment of epidemic substance abuse. It was not just happening behind the scenes, in dance clubs and high rise apartments; it was also rampant in comparative literature departments, design studios and corporate boardrooms. Postmodern practice embodied all this early on. It was ungrounded in terms of production and use value, and diagrammed the forces of mediation that separate production from consumption.

What began as a migration from structure to surface, then, dissipated entirely as postmodernism reached its precipitous end, leaving behind it only a sense of pervasive fragmentation within a vaguely defined cultural ether. This outcome is in some ways to be regretted, partly because it left postmodernism open to charges of being arch, ironic, or 'meta', as if its sole purpose were to place life in inverted quote marks. This is to seriously undersell the legacy of postmodernism, which was actually a power plant (in German, *Kraftwerk*) of innovation. We can recover the radical potential of postmodernism best by concentrating on the years when it was held between the modernist ideals of substance and subsequent notions of insubstantiality. In retrospect it is no surprise that the surface, as the interface between substance and context, was its preferred site of experimentation. The repertoire that postmodernists developed to articulate this surface is still in use today, partly because they were so prescient – Greiman's manually realized designs anticipate contemporary work produced on Photoshop, for example, while the dextrous turntable performances of Grandmaster Flash look ahead to digital mash-ups. But the postmodern surface also retains a hold on us for other reasons. Postmodernism did bequeath to the present a purposeful lack of closure, understandable in a period haunted by various kinds of endgame (not only the end of modernism, but the multiple deaths of the author, of painting, of analogue music). That can make its legacy seem to be one of paralysis. But, as this essay has hopefully demonstrated, we should rather draw the opposite conclusion. Alessandro Mendini once said that 'Superficiality has depth, if understood and accepted as the profound difficulty of human life.'[19] In staging their complicated manoeuvres on the surface of things, postmodernists were doing us a great service. They were putting design's discontents in plain view, right where we can see them.

Notes

1 *Postmodernism: Style and Subversion, 1970 to 1990* was staged at the Victoria and Albert Museum, London, from 24 September 2011 to 15 January 2012, and toured subsequently. The present essay is indebted to many colleagues who contributed to this project, especially my co-curator Jane Pavitt.

2 O. Welles (dir.), *The Lady from Shanghai* (Columbia Pictures, 1947).

3 C. Jencks, *The Language of Post-Modern Architecture* (London: Academy Editions, 1977).

4 For a survey of the topic see M. Golec and A. Vinegar, *Re-Learning from Las Vegas* (Minneapolis: University of Minnesota Press, 2009).

5 R. Martin, *Utopia's Ghost: Architecture and Postmodernism, Again* (Minneapolis: University of Minnesota Press, 2010), p. 122.

6 R. Venturi, *Complexity and Contradiction in Architecture* (New York: Museum of Modern Art Press, 1966).

7 C. Tulloch, 'Buffalo: style with intent', in G. Adamson and J. Pavitt (eds), *Postmodernism: Style and Subversion 1970 to 1990* (London: V&A Publications, 2011), pp. 182–185.

8 C. Jencks, 'Late-modern architecture' (1978), in *Late-Modern Architecture and Other Essays* (London: Academy Editions, 1980), p. 14.

9 Derrida posed the question as follows: 'Where does the frame take place. Does it take place. Where does it begin. Where does it end. What is its internal limit. Its external limit. And its surface between the two limits.' J. Derrida, *The Truth in Painting* (1978) (G. Bennington and I. McLeod trans.) (Chicago and London: University of Chicago Press, 1987), p. 61.

10 C. Rossi, 'Making Memphis: 'glue culture' and postmodern production strategies', in Adamson and Pavitt, *Postmodernism*, pp. 160–165.

11 W. Whitman, 'Song of Myself,' in *Leaves of Grass* (New York: 1855).

12 Author's interview with Peter Shire, 23 August 2009.

13 For the transforming drug trade in this period, see P. Gootenberg, 'The "Pre-Colombian" era of drug trafficking in the Americas: cocaine, 1945–1965', *The Americas*, 64:2 (2007), 133–176; P. Gootenberg, 'Between coca and cocaine: a century or more of U.S.-Peruvian drug paradoxes, 1850–1980', *Hispanic American Historical Review*, 83:1 (2003), 119–150; S. Topik, C. Marichal and Z. Frank (eds), *From Silver to Cocaine: Latin American Commodity Chains and the Building of a World Economy 1500–2000* (Durham, North Carolina: Duke University Press, 2006); S. B. Karch, *A Brief History of Cocaine: From Inca Monarchs to Cali Cartels* (Boca Raton, Florida: Taylor & Francis, 2006); and M. Taussig, *My Cocaine Museum* (Chicago: University of Chicago Press, 2004).

14 D. Banash, 'Intoxicating class: cocaine at the multiplex', *Postmodern Culture*, 12:1 (2001). He goes on to add, 'While marijuana films revel in satire, heroin films explore the complexities of self and self-destruction. Distinct from both are films about cocaine, which are almost always evocations of and reflections on the American dream itself, that is to say, on politics in the most practical and quotidian sense of the word.' http://pmc.iath.virginia.edu/issue. 901/12.1.r_banash.html.

15 Quoted in B. Bischofsberger (ed.), *Ettore Sottsass: Ceramics* (London: Thames and Hudson, 1995), p. 86.

16 'Ettore Sottsass Interviewed', *Memphis '82* (Milan: 1982), n.p.; reprinted as 'The Situation is Different', *Design*, 408 (December 1982), 38–39.

17 G. Van Sant (dir.), *Drugstore Cowboy* (Avenue Pictures, 1989).

18 B. Zevi, 'Pluralismo e Pop-architettura', *L'architettura*, 143 (September 1967), 283; Elton quoted in J. Lepore, 'Just the Facts Ma'am', *New Yorker*, 24 March 2008, p. 80.

19 Museo Alessi, *Designer Interviews: Alessandro Mendini*, Corraini (producer) (2008).

Index

amateur 51, 52, 114, 181, 182, 186, 189
Arlecchino 198
Auerbach, Frank 27–8
authenticity 5, 45, 47, 51, 53, 55–6, 69, 72, 73, 77, 78, 81, 90, 158, 207
  inauthenticity 86, 91, 130
authorship 187, 189, 205

Bachelard, Gaston 17, 19
Banner, Fiona 11
Barnett, Pennina 18, 20
Barrett Browning, Elizabeth 166–7, 177
Bauhaus 15
Belstaff 136, 141
Benjamin, Walter 55, 129, 187
binary 1, 2, 13, 162
Birmingham Back to Backs 80
boundary 13, 21, 98, 149, 151, 154, 161
Brandt, Bill 8
Brewster, David 114, 115
British Broadcasting Corporation (BBC) 134–6, 139, 143
Brodsworth Hall 79
Browne, Matilda 111–18

Calke Abbey 79, 80
candle light 117–18, 168
Cecula, Marek 10
Chalayan, Hussein 9
Chevreul, Michel-Eugène 115, 118
Codman House 79
Cold War 123, 126–7
Conan Doyle, Arthur 134, 135
conservation 66, 72–83
Corbusier, Le 15
cosmetics 13, 104, 126, 207
craftsmanship 37, 172, 187
  see also skill
Cumberbatch, Benedict 136
curtains 75, 76, 77, 78

deconstruction 19, 204
decoration 15, 16, 19, 36, 42, 84, 183
  in architecture 14, 84–5
  of ceramics 193, 195, 198
  interior 72, 73, 75, 76, 79, 175
  textiles 123, 130
  see also ornament
Deleuze, Gilles 20, 149–151, 162, 187
Derrida, Jacques 19, 183, 184, 190, 205
Dickens, Charles 167, 171, 172–3, 177
dirt 9, 16, 38–9, 52, 139
Douglas, Mary 16
Downey Jr, Robert 135, 137
Drayton Hall 79
Drouot auction house 45, 46–7, 52
Duchamp, Marcel 188, 190, 203

Eames, Charles and Ray 16
Eastlake, Charles 171, 174–5, 177
East Side Tenement Museum 80
embroidery 4, 19, 34–5, 37, 98–110, 123
  degradation of 106
  materials 98, 102, 103, 106
  stitches 98, 106
English Heritage 79
Englishwoman's Domestic Magazine 111–18
Erddig Hall 79
Evans, Caroline 19

fabric (textile) 9, 20, 34, 85, 99, 100, 103, 105, 106, 112
and American fashion 123–32
Carol Christian Poell 151–62
and ceramic printing 197–8
conservation of 76
in film costume 137, 142–3
see also textiles
façade 18, 20, 91, 160, 193–4, 201–3
fake 5, 51–5, 60, 69, 130
fetish 127, 134, 140–3
Fogarty, Anne 123–6, 128–130
advertising images 125–6, 128
fashion designs 123, 124, 130
fold 20, 141
Friedan, Betty 127

gas light 116–18, 168, 173–4
Gehry, Frank 204, 205
Godey's Lady's Book and Magazine (*Lady's Book*) 111–18
Gormley, Antony 27
graffiti 3, 22
Greenberg, Clement 182–3
Guattari, Félix 149–51, 162, 187

hapticity 27, 28, 30, 31, 125, 130, 148
haptic visuality 124
Hara, Kenya 26
Hebdige, Dick 22
Historic New England 79
Holmes, Sherlock 134–46

inrô 40–2

japanning 168

kaleidoscope 114–15
Kant, Emmanuel 183–4
Kenkô, Yoshida 38
kitsch 182–3
Kôami, Iyo 37
Kôami, Nagashige 37
Kristeva, Julia 20–1, 85

Lacan, Jacques 143, 176
lacquer 13, 34–42, 126, 147
Chinese 39
layers 36, 39
Makie 35–6
Negoro 35, 37–9, 41
Ritsuô (Ogawa Haritsu) 35, 40–2
and social class 37–9, 41
uses of 35–6
wear 37–8
Ladies' Cabinet 112, 113, 117
Leatherbarrow, David 15, 16, 19
Leddy, Thomas 1, 68
Loos, Adolf 14–15

McCardell, Claire 124, 131
McCracken, Grant 17, 39, 83, 85, 184
male gaze 98, 101, 102, 141
mandrake 197, 198
Margiela, Martin 19
Marinetti, Filippo Tommaso 30–1
Memphis design group 206, 208
Mendini, Alessandro 200, 205, 206, 210
mermaid 101, 103, 106
Mies van der Rohe, Ludwig 15
Miller, Daniel 21–2, 30, 131
modernism 15–16, 84–5, 182, 195, 202–6
modesty 103, 105, 123, 152
Moffatt-Ladd House 77–8
Mostafavi, Mohsen 15, 16, 19

nakedness 98, 99, 100, 102–3, 104, 143, 159
see also nude
National Trust 74, 75, 76, 78, 79, 80
National Trust for Historic Preservation 78
Newton, Isaac 115
nude 158, 182
see also nakedness

Ogawa Haritsu 40
see also lacquer, Ritsuô (Ogawa Haritsu)
ornament 2, 14–15, 50, 174–6, 195, 202, 206
see also decoration

Pajaczkowska, Claire 18, 141
parergon 183, 184, 190
patina 17, 38, 39, 45–59, 68–70, 73, 74, 78, 80–1, 86, 137–9, 184
pattern (decorative) 14, 15, 39, 40, 63, 69, 152, 153, 206
pattern (design) 99, 152, 155–7, 160
photography 8, 51, 185, 188,
  fashion 124, 129, 131
Picasso, Pablo 204
Poe, Edgar Allan 167, 173–4, 177
polish 7, 8, 11, 13, 16, 17, 36, 45, 51, 167, 168,
  French 14, 169, 171, 175
Pollock, Jackson 183, 190
poncif 187
Postmodernism 16, 19, 93, 200–211
print 35, 84, 93, 206
  ceramics 193–8
  photographic 8
  source for embroidery designs 99, 100, 101
  textiles 15, 19

Rauschenberg, Robert 204
readymade 11, 188, 190
Ritchie, Guy 134–5, 137–8
Rothko, Mark 184
Ruskin, John 167, 174, 177, 195

Scott Brown, Denise 16, 202, 203, 209
skill 2, 14, 15, 29, 42, 123
  cabinet-making 5, 169, 176
  embroidery 19, 98, 99
  paint-by-number 185, 186, 188, 189, 190
  weaving 66, 67, 69–71
  see also craftsmanship
skin 18–20, 22, 126, 128, 131, 154–5, 157–60
  complexion 103–4
  embroidery of 98–110
  and touch 26–7
Strawbery Banke 81
Studio Alchymia 205, 206
Sturbridge Village 81
Sudo, Reiko 29
sumptuary laws 34, 37
Sunnycroft 79, 80

Tactillism, Manifesto of 30
Tanizaki, Jun'ichirô 38
tea ceremony 10, 39, 40
Tennyson, Alfred (Lord) 166
textiles 13, 9, 15–16, 18–20, 28, 29, 60, 61, 85, 140, 142–3, 151–2, 193, 197, 198
  ageing of 60–71
  and American fashion 125
  colour and 111–19
  conservation of 72–81
  see also fabric (textile)
Tokugawa period (1615–1868) 36, 37
Tyntesfield 76, 79

Uppark 75–6, 78–9
Utamaro 35

varnish 13, 14, 21, 45, 51, 52, 53, 168–71
veneer 147, 168, 172, 173, 174, 202
Venturi, Robert 16, 202–3, 209
Victoria and Albert Museum 4, 41, 195
Vigarello, Georges 18
Von Goethe, Johann Wolfgang 115

wallpaper 72, 73, 76, 77, 85, 183
Warhol, Andy 189
Wigley, Mark 15
Wilde, Oscar 18
Woolf, Virginia 167
Worth, Charles Frederick 116